Soft Computing in Engineering

T0179254

Soft Computing in Engineering

Jamshid Ghaboussi

CRC Press
Taylor & Francis Group
Boca Raton London New York

CRC Press is an imprint of the
Taylor & Francis Group, an **informa** business
A SPON PRESS BOOK

CRC Press
Taylor & Francis Group
6000 Broken Sound Parkway NW, Suite 300
Boca Raton, FL 33487-2742

First issued in paperback 2020

© 2018 by Taylor & Francis Group, LLC
CRC Press is an imprint of Taylor & Francis Group, an Informa business

No claim to original U.S. Government works

ISBN-13: 978-1-4987-4567-3 (hbk)
ISBN-13: 978-0-367-65726-0 (pbk)

Visit the Taylor & Francis Web site at
http://www.taylorandfrancis.com

and the CRC Press Web site at
http://www.crcpress.com

To Jennifer

Contents

Preface

The primary purpose of this book is to present a new viewpoint in engineering applications of soft computing tools such as neural networks and genetic algorithm. The author aims to present a case for treating these soft computing tools differently than the other traditional mathematically based engineering problem-solving methods. All the engineering problem-solving methods are based on classical mechanics and mathematics. It is therefore natural to consider any new method as another mathematically based method. This book intends to describe and emphasize the fundamental differences between the biologically inspired soft computing methods and the traditional mathematically based methods. It is hoped that this will encourage more innovative uses of these new soft computing tools and that it will lead to exploiting their full potential.

The biological origin of the soft computing methods has been emphasized intentionally. Biological systems in nature, such as insects, insect colonies, animals, humans, and brains and immune systems, to name a few, have evolved to be very effective methods for solving highly complex and difficult problems. Some of the problem-solving strategies used in nature can also be deployed in solving engineering problems, especially those problems that are very difficult, if not impossible, to solve with the traditional mathematically based methods. Included in this class is the large number of inverse engineering problems that we either do not attempt to solve, or we do not solve them directly as inverse problems.

On occasions throughout this book we will refer to the problem-solving capabilities of the human brain that is a highly complex and sophisticated computer. Human brains are able to solve very difficult problems in the day-to-day activities of human beings. The same observation also applies to animal brains. The structure and operation of brains are very different than our computers. Brains have evolved into massively parallel computers to address the needs of the organisms in nature. It is tempting to say that neural networks are based on the structure of the human brain, except that we know very little about the internal workings of the human brain. Nevertheless, there is some truth in the above-mentioned statement. Neural networks are also massively parallel systems and, in a crude and highly simplified manner, they mimic the function of the neurons and synapses. Even our limited knowledge of the internal workings of the brain suggests novel and potentially effective methods for solving some difficult engineering problems, most of which are inverse problems. It is the hope of the author to encourage the reader to seriously consider the engineering application of the problem-solving strategies utilized in nature and by our brains.

We also discuss computational methods that emulate the natural Darwinian evolution. This class of very effective methods is called genetic algorithm or simply evolutionary methods. These methods start by developing a population whose individuals represent an engineering problem. In simulating a number of generations of that population, some of the basic elements of the natural evolution are used; they include genetic recombination, natural selection, and random mutation. Effectiveness and robustness of these methods are based on the fact that they mimic the natural evolution and we can observe the spectacular success of natural evolution in producing highly complex organisms. At its most basic level the evolutionary methods are considered as effective optimization methods. However, I believe that these methods have far greater potential in tackling some of the currently intractable problems in design and other difficult inverse problems in engineering.

In the introductory chapter, I present the main thesis of this book and discuss how we do engineering and how nature solves problems, and what are the fundamental differences between these two approaches to problem-solving. This is followed by four chapters on neural networks and three chapters on genetic algorithm. The basics of neural networks and genetic algorithm are presented in Chapters 2 and 6 along with simple examples illustrating how they work and the limits of their utility. The other chapters present extensive examples of the application of computational intelligence methods based on neural networks and genetic algorithm to various classes of engineering problems. In these chapters, I have presented many examples of the engineering applications of neural networks and genetic algorithm to emphasis and further develop the main thesis of the book presented in the introductory chapter. Necessarily, these examples come from my own research, since I am most familiar with them and can explain them well. Consequently, the examples are from my own field of expertise, which is civil engineering. I believe that the principles that they are intended to illustrate are independent of their field of application. These examples are intended to convey the viewpoint that neural networks and genetic algorithm are fundamentally different than the other traditional mathematically based methods and, as such, they can be used in innovative ways to exploit their full potential. Some new and advanced forms of the neural networks and genetic algorithm, which have been developed in my own research group, have been presented throughout this book. Again, these advanced forms are an important component of making the main point of the book; that is, biologically inspired soft computing tools of the computational intelligence can be used effectively to solve difficult engineering problems that cannot be solved with the conventional mathematically based engineering methodology.

This book is intended for a reasonably wide audience. A minimum level of engineering knowledge, equivalent to a senior level, or the first engineering degree, is assumed. On the one hand, those who have no knowledge of the subject matter, and they are curious and wish to learn more about the engineering applications of these computational tools, would find that this book will satisfy their needs. On the other hand, those who have knowledge of the subject matter, and perhaps are even doing research in this field, will find that the book provides new and, hopefully, interesting points of view on the engineering applications of the soft computing tools. This book can also serve as a text for a senior level or the first year graduate level engineering course on neural networks, genetic algorithm, and their engineering applications. Finally, the reader without a technical background can also benefit from gaining some understanding of the interesting relationship between the nature's problem-solving strategies and the engineering methodology.

The author's research over the past 30 years forms the basis for this book. Over this period, I have collaborated with colleagues and have had the benefit of working with many excellent doctoral students. The examples used throughout this book are mostly based on the doctoral dissertation researches of my former students. I am greatly indebted to them, although all the errors, omissions, and short comings of this book are mine alone. I express my gratitude and appreciation for the opportunity for collaborating on research projects with my colleagues: Professors David Pecknold, Youssef Hashash, and Michael Insana of University of Illinois at Urbana-Champaign; Professor Amr Elnashai, currently vice chancellor for Research and Technology Transfer at the

University of Houston; and Professor Poul Lade, currently at George Mason University. I am grateful to Professor Yeong-Bin Yang and the National Taiwan University for giving me an opportunity to spend a sabbatical semester in Taiwan. I started writing this book during that sabbatical leave.

Jamshid Ghaboussi

Author

Jamshid Ghaboussi is an emeritus professor in civil and environmental engineering at the University of Illinois at Urbana-Champaign. He received his doctoral degree from the University of California at Berkeley. He has more than 40 years of teaching and research experience in computational mechanics and soft computing with applications in structural engineering, geomechanics, and biomedical engineering. He has published extensively in these areas and is the inventor of five patents, mainly in the application of soft computing and computational mechanics. He is the coauthor of books *Numerical Methods in Computational Mechanics* (CRC Press) and *Nonlinear Computational Solid Mechanics* (CRC Press). In recent years he has been conducting research on complex systems and has coauthored a book on *Understanding Systems: A Grand Challenge for 21st Century Engineering* (World Scientific Publishing).

Chapter 1

Soft computing

1.1 INTRODUCTION

The term soft computing refers to a class of computational methods that are inspired by how biological systems operate in nature. We can think of them as problem-solving *computational methods* employed in nature by the biological systems. Soft computing methods are also referred to as *computational intelligence methods*. Included in this class of computational methods are artificial neural networks, genetic algorithm, fuzzy logic, and swarm intelligence. Neural networks are roughly modeled after the structure and the operation of brains and nervous systems in humans and animals. Genetic algorithm is modeled after the natural Darwinian evolution, fuzzy logic uses the linguistic approaches to problem-solving, and swarm intelligence is based on the decentralized and self-organized behavior of systems such as ant colonies. In this book, we will mainly concentrate on neural networks and genetic algorithm.

Neural networks and genetic algorithm were introduced several decades ago. It is only during the past two or three decades that they have been increasingly used in all fields of engineering, as well as nonengineering fields such as biomedicine and business. As expected, with the introduction of any new methods, the earlier applications of the neural networks and genetic algorithm can be reasonably classified as routine. This mainly means that they were initially treated as another computational tool and were used similar to other mathematically based hard computing methods. However, we will see in this book that the soft computing methods are fundamentally different from the mathematically based hard computing methods and have potential to address some of the most difficult and currently intractable engineering problems. These methods have matured to a point that it is plausible to consider their applications to some of these difficult and currently intractable engineering problems. We are at the next stage of the development and application of these methods in engineering problems, especially in engineering mechanics. The successful application of these methods to difficult problems in engineering mechanic will not only require a thorough understanding of the fundamentals of the soft computing methods and how they differ from mathematically based hard computing methods, but also an understanding of how nature uses them to solve problems.

The emphasis on the biological origin of the soft computational methods is important. Understanding their biological origin, and how they work in nature and what types of problems are solved by these methods, is of critical importance in recognizing their full potential in solving engineering problems. This is especially important when we consider the application of the soft computing methods to engineering problems that either cannot be solved or are difficult to be solved with the conventional mathematically based methods. Many of the problems that humans and animals routinely solve using their brains, which can be considered as biological computers, have counterparts in engineering, and a vast majority of these problems are intractable with the mathematically based engineering methodology. Similarly, evolution promises to be an effective model for solving many difficult engineering problems, especially in engineering design. Understanding of how animals and humans and evolutionary process solve problems in nature will help the development of effective and innovative applications of soft computing methods leading to new and creative ways of using them in solving difficult engineering problems.

Simple and routine application is to use neural networks and genetic algorithm as an alternative method to solve engineering problems that can also be solved with the conventional mathematically based methods. It is far more important to use these methods to solve problems that cannot be solved with the current methods. In this book, I present neural networks and genetic algorithm as new computational methods that fundamentally differ from the mathematically based engineering methods and discuss these fundamental differences. Throughout this book, I have emphasized using the unique capabilities of these methods to develop innovative approaches to solving problems in engineering that either cannot be solved or are difficult to be solved with conventional mathematically based hard computing methods.

1.2 HARD COMPUTING AND SOFT COMPUTING METHODS

Neural networks and genetic algorithm are referred to as "soft computing methods." The term "soft computing" has connotations of imprecision tolerance, and this is precisely what differentiates them from the "hard computing methods" that are overly reliant on the computational precision. Almost all the engineering computations performed on the current computers belong to the class of hard computing methods.

In the field of computational mechanics, large-scale computations are performed using finite-element method or finite-difference method to simulate the response of the mechanical systems to external stimuli such as loads and temperature change. Nonlinear dynamic finite-element analyses are routinely performed on very large discrete systems involving tens of thousands of finite elements. For example, large-scale analyses are performed to study and evaluate the structural response of solids and structures, or to simulate the crash of automobiles, or to study the aerodynamics of aircraft in flight, or to study the behavior of biological systems in biomedicine. Similar large-scale computations are also performed in computational fluid dynamics using either finite-element method or finite-difference method.

The most fundamental characteristic that all these methods—as well as many similar types of computations using other modeling techniques—share is that they are hard computing methods, in contrast to the soft computing methods, which is the subject of this book. We refer to them as hard computing methods to emphasize their reliance on the precision. The source of the high level of precision in the hard computing methods is the fact that they are based on mathematics which is precise.

As illustrated in Figure 1.1, hard computing methods require precise inputs and they give precise outputs, and the computation is done with a high degree of precision. Precision in most cases in computational mechanics implies "exactness" to within the round-off error. In a typical finite-element analysis, the input consists of the geometry of the structural system, boundary conditions, material properties, and the external stimuli, such as loads.

In most engineering problems, these quantities cannot be determined precisely, and there is some degree of uncertainty, error, scatter, and noise in the data. This is especially true in material properties. Moreover, the assumption that the same material properties can be used to characterize the material behavior over regions of a structural system is often not accurate

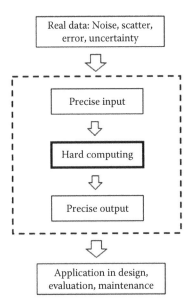

Figure 1.1 Illustration of the precision-dependent hard computing in engineering application.

and usually there are some spatial variations in material properties. However, the finite-element analysis requires complete and precise input, perhaps up to 14 digits if the analysis is performed in double precision. Even though we may not regard the input as precise, the computational method treats the input as precise. The output of the hard computing methods is also precise or exact to within the round-off error, although the high degree of precision is not required for most engineering applications.

On the other hand, as we will see throughout this book, the biologically inspired soft computing methods are inherently imprecision tolerant. The imprecision tolerance of the soft computing methods is their key feature, and it plays a central role in defining the main characteristics of these methods and their potential effectiveness in solving many difficult and currently intractable problems in engineering.

It is useful to think of the soft computing methods as fault tolerant computing, both external and internal fault tolerance. External fault tolerance implies more than the capability to deal with imprecise input data for the problem. It means that the soft computing methods have the capability of dealing with an imprecise, noisy, and incomplete data within some limits. Similarly, internal fault tolerance means that the soft computing methods can suffer certain amount of internal damage and still perform reasonably satisfactorily. In the computational form of the soft computing methods, internal fault tolerance implies some degree of insensitivity to the algorithmic detail of the methodology.

To clarify the imprecision tolerance and fault tolerance of the soft computing methods, we need to go back to their biological origin and examine the biological systems that inspired these methods. Humans and animals live, operate, and have to survive in environments that they live in.

The reader may feel unsatisfied with the lack of rigor in the preceding discussion and the use of terminology without rigorous definition. We mention these issues at this point just to introduce the subject. Later in this book, we will devote more space to in-depth discussion of these issues.

1.3 MATHEMATICALLY BASED ENGINEERING
PROBLEM-SOLVING METHODOLOGY

The subject of this section may appear somewhat out of place for this book. The relevance of this subject to the central core of the thesis of this book will become apparent later. At this point, it suffices to say that as we are studying the engineering applications of biologically inspired soft computing methods, it is important to clearly understand the basic differences between these methods and the mathematically based engineering methodology. To accomplish this, we need to first explore the basic properties of the engineering methodologies that the engineers normally do not think about as they are so fundamental that they are often taken for granted. However, a reexamination of those fundamental aspects of engineering methodologies are the key to understanding the main differences between the biologically inspired soft computing methods and the mathematically based engineering methodology.

The main mission of engineering is to create and maintain new objects, structures, and systems. A whole spectrum of objects and structures have been and are being created by human activity. They range from simple objects such as furniture to more complex objects and structures such as computer chips, space craft, and long-span bridges. The carpentry that produces the furniture is not considered engineering, whereas computer chips and long-span bridges are. In moving from simple to complex, there is a somewhat hazy line separating trades from engineering. That line has also changed with time. Objects have been made and structures have been built throughout the human history. For example, the invention of the wheel was a milestone in the human development. However, it is not considered engineering. In fact, the word engineer is only a few hundred years old and engineering as a profession is barely two hundred years old. The modern profession of engineering differs from the other trades and practices of the past as it is based on the scientific principles. Creating complex new computer chips and building long-span bridges definitely requires knowledge of mathematics, physics, and scientific principles. However, the relationship between science and engineering is complex, and its thorough examination would require a separate volume. We will only attempt to briefly discuss this subject to lay the ground work for development of some important ideas to follow.

Science deals with the discovery of new knowledge. It studies the natural phenomena to discover the underlying unifying principles that explain the phenomena. As engineering deals with the problem-solving methods in the real world, all disciplines of science are relevant to engineering to varying degrees. The most basics of the scientific disciplines from engineering point of view are physics and mechanics that are based on the mathematical principles. The scientific disciplines of chemistry, biology, and geology are to lesser extent dependent on mathematics. This relationship with mathematics is very important for us. Physics and mechanics seek universal and unifying principles expressible in mathematics. It is often said that the hierarchy of scientific disciplines form a pyramid. At the base of the pyramid is mathematics and above that are physics and mechanics. Each level in the pyramid depends on all the scientific disciplines below it. Above physics, is chemistry, and above that are biology and geology and so on.

One can observe that as we move up in the scientific pyramid, random variability, uncertainty and scatter increases and the natural phenomena are less and less expressible by the strict mathematical constructs. The motion of a body may be expressed mathematically with a high degree of accuracy. The same cannot be said about the study of occurrence of earthquakes and their underlying mechanisms.

Engineering deals with the real-world problems. Science can justifiably strip away the uncertainty, variability, scatter, and noise to discover the underlying unifying principles; this approach is known as *reductionism*. Engineering tasks require not only the use of the scientific principles, but also taking into consideration the natural variability and scatter, and accounting for uncertainty and lack of knowledge. Scientists can choose what phenomena to study but engineering tasks must be performed regardless of the availability of complete knowledge. Engineers often have to deal with some degree of uncertainty resulting from the lack of complete scientific knowledge.

It is reasonable to state that mathematically based methods have the following three properties:

1. Precision
2. Universality
3. Functional uniqueness

We normally do not think about these three fundamental properties of mathematically based problem-solving methods used in engineering. We will discuss these in detail as they form the basis of the fundamental differences between the hard computing and soft computing methods.

Precision: Of course, precision in mathematically based methods is self-evident and does not require much elaboration. We need to point out that to address uncertainty, often artificial randomness is introduced in the branch of mathematics known as stochastic.

Universality: By universality we mean that mathematical functions, such as sin(x), are valid for all the possible values of x, from $-\infty$ to $+\infty$. Mathematical models of physical phenomena share this universal property. For example, when we assume that a material behavior is linearly elastic, the equation that we write implies that the material behaves linearly elastic for all the possible values of strain from $-\infty$ to $+\infty$. In reality, we know that this is not true–the material may behave approximately linearly elastic for a small range of strain around the origin, but the equation that we use to represent this local behavior is universal.

Functional uniqueness: Here we observe that mathematical functions are unique. For example, there is only one sin(x).

These three properties of mathematics that transfer to the mathematically based hard computing methods used in engineering problems are obvious, but we usually do not think about them, because there are no alternatives. However, soft computing methods are fundamentally different; they are based on nature's problem-solving strategies and, as such, they do not share the fundamental properties of the mathematically based hard computing methods.

1.4 PROBLEM-SOLVING IN NATURE

We can think of human and animal brains as biological computers, and they are constantly solving problems. For example, when a tiger sees its prey, it needs to know how far it is. However, it does not need to know the distance to its prey precisely; approximate distance is sufficient. The task of computing the distance is highly complex—it requires receiving signals from the eyes, coordinating them, and using the past experience to determine the approximate distance to its prey. The animal's brain performs this computational task and provides the imprecise, but sufficiently useful, value in real time.

One of the important classes of problems in nature is recognition, such as voice recognition, face recognition, recognizing the prey, recognizing danger, and so on. Our brains are capable of recognizing voices of the people we know. Similarly, we do recognize the faces of the people we know. These functions are performed with a high degree of robustness and imprecision tolerance. For example, we would recognize the voice of someone we know if we talk to them on the phone or even when they have a cold and their voice is hoarse. Similarly, we would recognize the face of someone we know, even if it is partially covered. Another class of problems in nature is control; our brains control our limbs or control the functioning of internal organs.

These functions share the following important characteristics:

1. Imprecision tolerance
2. Nonuniversality
3. Functional nonuniqueness

Imprecision tolerance: We have briefly discussed the imprecision tolerance: The predator needs to know the approximate distance to its prey; and face and voice recognition does not need precise input; it can be accomplished with somewhat modified and/or incomplete input. The computational tasks are performed by brains in a robust and imprecision tolerant manner.

Nonuniversality of these biological computational tasks means that they are performed within a relevant range of the input. For example, a predator does not need to know the distance to its prey for all the possible distances from 0 to ∞. It only needs to know the distance within a reasonable range. Similarly, we are not capable of recognizing all the possible voices and faces. Studies have shown that humans are capable of recognizing about 150 to 200 faces. This correlates with the number of individuals who lived in early human communities in which recognition was a matter of survival. One of the contributing factors to the growth of human brain sizes is thought to be social interactions that involved voice and face recognition in the groups that early humans interacted with socially.

Functional nonuniqueness: This means that these biological computational tasks discussed earlier could be performed to within reasonable accuracy by different brains that have different internal structures, unlike mathematical functions that are unique.

Computational capabilities in biological systems have evolved to meet certain requirements that were essential to their survival. All these computational tasks needed to be performed in real time. A predator needs to recognize its prey and to know the distance to its prey now, in real time. Similarly, we need to recognize voices and faces in real time. Similarly, our brains need to control our organs and limbs in real time. Massively parallel structure gives brains the capability of performing these tasks in real time, unlike our sequential computers that are much slower.

Another important characteristic of these computational tasks is that they are all inverse problems. Almost all problems encountered by biological systems in nature are inverse problems.

All computational tasks can be classified into either forward problems or inverse problems. To clarify this point, we need to examine a typical computational task that involves simulating the response of a system subjected to a stimulus. In computational simulations, we have three entities: the model of the system; an input (stimulus); and an output (response), as shown in Figure 1.2. In the forward problems, the model of the system and the input are known, and the output is computed. In the inverse problems, the model of the system and the output are known, and the input that generated that output needs to be computed (Figure 1.3). A second type of inverse problem, which will be discussed in Chapters 4 and 8, is when the input and the output are known, and the model of the system needs to be determined.

Figure 1.2 A typical system with input and output.

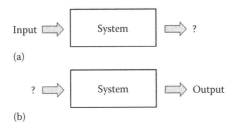

(a)

(b)

Figure 1.3 (a) A typical forward problem, in which the input is known and the output needs to be is determined. (b) A typical inverse problem in which the output is known, and the input needs to be determined.

In the case of recognition, such as voice recognition and face recognition, the forward problems are when we know an individual and hear his/her voice and see his/her face. The inverse problem is when we hear a voice or see a face and have to identify the person. In the case of control, the forward problem is to send a known signal to a limb or an organ and observe its response. The inverse problem is to determine and send the input signal that would produce the desired response in a limb or an organ.

We can see that biological computational methods have evolved to solve the inverse problems in real time. Biologically inspired soft computing methods that will be discussed in this book inherit these capabilities from the biological computational systems. We will see later that neural networks are massively parallel systems, and they can be used in the forward problems as well as in the inverse problems. We will give examples of application of neural networks and genetic algorithm in inverse engineering problems.

In the next section, we will discuss some of the engineering inverse problems. Mathematically based engineering problem-solving methods are only suitable for forward problems. As a result, either we ignore engineering inverse problems, or we transform them to iterative use of forward problems.

The importance of understanding of the fundamental differences between the mathematically based hard computing methods and biologically inspired soft computing methods cannot be overemphasized. To exploit the full potential of the soft computing methods in engineering applications, we need to be aware of these differences. The general tendency at first is to the use the soft computing methods the same way as the mathematically based hard computing methods are used in solving the forward problems. Throughout this book, we will emphasize the full potential of soft computing methods and present case studies, from my own research, to illustrate these points.

1.5 DIRECT AND INVERSE ENGINEERING PROBLEMS

We will start by examining typical engineering computational tasks. The objective of most typical engineering computational tasks is to determine the response of a system to external and/or internal stimuli. We can think of the stimuli as the input and the system's response as the output. In this case, we have three entities as shown in Figure 1.2: the system, the input

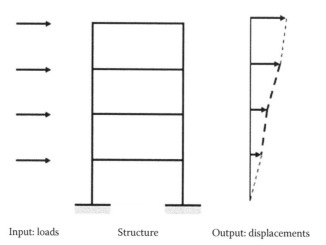

Input: loads   Structure   Output: displacements

Figure 1.4 A structural system with input loads and output displacements.

to the system, and the system's output. Of course, in a computational task, we are dealing with computational model of the system. While the system is fixed, we define what the input is and what the output is.

As an example, we will consider a structural system, such as the steel frame of a multi-story building, shown in Figure 1.4. Structural engineers routinely analyze the computational models of such structural systems. At the structural degrees of freedom, there are two complimentary variables: forces and displacements. If we designate forces as input, then the displacements will be the output. We can also apply the displacements as input and determine the forces that caused those displacements as output. In Figure 1.4, forces are assumed to be the input and displacements are designated as the output.

In a forward analysis, the computational model of the structural system and the inputs are known and the output of the system is computed. This is the typical computational task in computational mechanics. In the case of structural analysis, we have a numerical model of the structural system, and the applied loads are known and we perform a forward analysis to determine the structural displacements, as shown in Figure 1.5.

In the inverse analysis, we normally know the output of the system. Then, we have the input to the system and numerical model of the system—one of these two will be known and the other to be determined. Therefore, there are two types of inverse problems. In the *type one inverse problem*, as shown in Figure 1.6a for the structural system, we have the computational model of the structural system and we have measured the displacements of the structure, and in the inverse analysis we need to determine the forces that caused those displacements.

The *type two inverse problem*, as shown in Figure 1.6b, is determined when we know the input to the system and have measured the response of the system (output). The objective of the inverse analysis is to determine the computational model of system. In the case of the structural analysis, we have the applied forces and have measured the displacement response of structural system and need to determine the computational model of the structural system.

In computational mechanics, all the finite-element analyses are in the form of forward analyses. We can state that the mathematically based computational methods are only suitable for analysis of forward problems. They cannot be directly used for solving inverse problems. It is for this reason that in engineering problem solving, we only use forward analyses and seldom think of directly solving inverse problems.

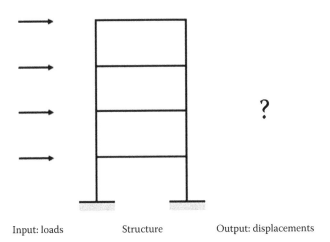

Input: loads   Structure   Output: displacements

Figure 1.5 A typical forward analysis in which the model of the structural system and the input loads are known, and the output displacements are the unknowns to be determined.

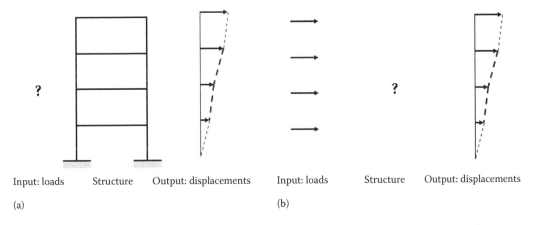

Input: loads   Structure   Output: displacements   Input: loads   Structure   Output: displacements

(a)   (b)

Figure 1.6 Two types of inverse problems in structural analysis: (a) Structural model is known and displacements have been measured, the forces that cause those displacements are the unknowns to be determined; (b) applied forces are known, and displacements have been measured, the structural model is the unknown to be determined.

There are many examples of type two inverse problems in engineering, and we will discuss a few of them here. Material modeling from the results of material experiments is one example. In material tests, a sample of the material is subjected to known forces or stresses and the response of the sample in the form of displacements and/or strains are measured. The objective is to determine the material model. This is seldom treated as an inverse problem, and it is not directly solved as an inverse problem. Usually, a mathematical model of the material behavior that satisfies the laws mechanic is assumed on the basis of the observed material behavior, and its parameters are determined by trial-and-error or some optimization method. In Chapter 5, we will present a direct method of solving this inverse problem using artificial neural networks.

Another example of type two inverse problem is condition monitoring of structural systems. The response of the structural system, such as a bridge, to known applied forces is measured. Either known forces are applied in an experiment, or the forces exerted on the structure, such as traffic loads on bridges, are measured. The objective of condition monitoring is to determine some aspect of the structural model, such as fracture or other types of internal damage that may lead to structural damage or failure. This is a subject of active research and many investigators are using different methods. Most of the approaches involve optimization methods to minimize the difference between a response computed in a forward analysis and the measured response. In Chapter 7, we will present a direct solution of the condition monitoring using a special type of genetic algorithm.

In medicine, biomedical imaging is inherently a type two inverse problem. Soft tissue is stimulated, and its measured response is used to generate an image. There are many examples of biomedical imaging, such as ultrasound, X-ray, magnetic resonance imaging, and computed tomography scan. Many imaging methods generate qualitative images that give relative properties of the soft tissue. Elastography is a method of generating quantitative images of the mechanical properties of the soft tissue that may contain important diagnostic and other medical information. Elastography is the subject of extensive current research. In Chapter 5, we will briefly discuss a soft computing method of determining the mechanical properties of soft tissue by directly solving the inverse problem with artificial neural networks.

Earlier, we mentioned that the mathematically based methods are not suitable for direct solution of the inverse problems, especially for type two inverse problems. Most applications of mathematically based methods to inverse problems use some form of optimization method with repeated application of the forward analysis. Unlike forward problems that have unique solutions in deterministic methods, inverse problems may not have unique solutions. The nonuniqueness is magnified with the lack of sufficient information in the form of measurements and the accuracy of those measurements.

Biological systems have evolved methods of solving inverse problems uniquely. Biologically inspired soft computing methods are potentially capable of solving inverse problems uniquely. The methods presented in the later chapters arrive at unique solution for the inverse problems.

We will briefly discuss how nature solves inverse problems uniquely. But before doing so, we need to discuss another class of inverse problems in engineering and in nature. Engineering design is a type two inverse problem, as illustrated in Figure 1.7. Normally design is not directly solved as an inverse problem. The main approach to engineering design is partly creative and partly gradual utilization of the past experience.

Design in nature occurs through evolution. Genetic algorithm that is based on the natural evolution can be used in engineering design, as we will demonstrate through some examples in Chapter 8. However, most of the current applications of genetic algorithm are in optimization. In the next section, we will discuss how nature solves inverse problems, and how those methods carry over into the soft computing methods.

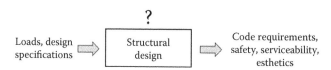

Figure 1.7 Engineering design as a type two inverse problem.

1.6 ORDER AND REDUCTION IN DISORDER

We can think of evolution as a form of problem-solving in nature—it is nature's equivalent of engineering design. In engineering design, we have some specification and requirements. We are also constrained by our past experience and limited scientific knowledge of behavior of the systems. As such, engineering design becomes a search in a fairly small finite-dimensional vector space, whereas true design can be a search in an infinite-dimensional vector space. Evolution in nature is a search in infinite-dimensional vector space, with few, if any constraints.

Nevertheless, both engineering design and evolution in nature have some important common features. Both of them are searches in vector spaces. Next, we need to elaborate on what we mean by *search*. In both cases, search in vector spaces leads to reduction in disorder. Engineering designs have far higher order than the materials used to create them. Evolution has also led to plants and animals with much higher levels of order than the materials around them. Evolution started at the primordial soup with a high level of disorder and led to beings with increasingly higher levels of order. We can also think of order as *information*. Engineering designs have more information in them than the materials used to create them. Plants and animals also have far more information than the material around them.

Reduction in disorder and increase in the information content are the fundamentals in design and evolution. Genetic algorithm that is covered in Chapters 6–8 is modeled after natural evolution, although in a much simplified way. We can think of genetic algorithm as reducing disorder and increasing information content. Genetic algorithm is mostly used in simple optimization problems. In Chapter 7, we will see that we can formulate genetic algorithm to get closer to natural evolution and thereby increasing the dimensionality of the vector space in which the search takes place. Of course, the true design takes place in infinite-dimensional vector space in which the only objective is to reduce disorder in meeting the design objectives. We will present some examples of application to genetic algorithm in design in Chapter 8.

We should point out that almost all problem-solving methods in nature lead to reduction in disorder, including learning. This is relevant as neural networks learn information from the training data. Training neural networks lead to higher information content and reduction in disorder.

Reduction in disorder and increase in information content are the fundamental mechanism behind both neural networks and genetic algorithm that are the main soft computing methods covered in this book. This is the fundamental similarity between the soft computing and problem-solving methods in nature.

1.7 SUMMARY AND DISCUSSION

In this chapter, we have emphasized that soft computing methods are based on the problem-solving methods in nature. In that sense they are fundamentally different from the mathematically based problem-solving methods in engineering. Animal and human brains have evolved to solve problems with important survival implications in real time, within range of interest and imprecision and fault tolerance. Almost all these problems are inverse problems. Soft computing tools of neural networks and genetic algorithm that are covered in the book inherit their properties from the problem-solving methods in nature.

Mathematically based engineering problem-solving methods are precise, lack fault tolerance, and are mainly suitable for solving forward problems. We discussed the fundamental differences between mathematically based engineering problem-solving methods and soft computing methods. Mathematically based engineering problem-solving methods are precise, universal, and functionally unique. Soft computing methods are imprecise, nonuniversal, and functionally nonunique. These properties make soft computing methods suitable for solving inverse problems.

We discussed the inverse problems in engineering, including engineering design. As mathematically based engineering problem-solving methods are only suitable for solving forward problems, most inverse problems in engineering are not solved or are solved as forward problems.

Neural networks and genetic algorithm are roughly modeled after human brains and natural evolution in a highly simplified manner. One important strategy in improving the effectiveness of these methods is to move them closer to their equivalents in nature. We know that human brain is a highly effective computer, and we can judge the power of natural evolution by the highly sophisticated being that has evolved. We will give an example in Chapter 7 of a genetic algorithm that has taken the first step in that direction.

Chapter 2

Neural networks

2.1 INTRODUCTION

Neural networks are roughly patterned on the operation and structure of the brains and nervous systems of humans and animals. Their origins go back to more than five decades in attempts by psychologist and neurobiologist to develop computer models to gain more understanding of the internal working of human brain. Advances in these fields accelerated with the improvements in the speed and capacity of the computers. Physical scientists and engineers are the latecomers to this field, and the initial attempts to apply neural networks in engineering problems started in the early 1960s. Over the past two decades, the engineering applications of neural networks have accelerated and now they are used widely in many fields of engineering.

As stated earlier, the primary objective of this book is to lay the ground work for viewing neural networks and genetic algorithm as new and fundamentally different computational tools and to make the case for innovative methods of using these tools to realize their full potential. In the current chapter, we cover the basics of neural network to enable a beginner in this field to be able to follow the rest of the chapters in this book. We will cover enough material on the subject to bring the reader close to the state of the art. Some of the more advanced topics on neural network will be covered in the next few chapters.

We will mainly cover multi layer-feed-forward neural networks that have been used in the example applications presented in the next three chapters. We will also briefly describe Hopfield nets that are dynamical systems. In the discussion at the end of the chapter, we will mention and briefly discuss a few other types of neural networks. Multilayer feedforward neural networks are the most widely used and the most useful for engineering applications, especially in the field engineering mechanics that is the main topic of this book. We refer the reader to the more typical books on neural networks for more comprehensive coverage of the different types of neural networks.

2.2 ARTIFICIAL NEURONS

Artificial neural networks are constructed as an assemblage of artificial neurons that are roughly modeled after the biological neurons in the brains and nervous system of humans and animals. We start this section with a brief introduction to the structure and operation of the biological neurons. However, a warning is needed before we proceed with this introduction; what we present is a highly simplified picture of the biological neurons. Biological neurons have evolved over the past millennia. Much remains to be discovered about the highly complex structure and operation of biological neurons. The simplified view of the neurons presented here is justified as the current generation of neural networks uses artificial neurons that conform to this simplified version of neurons.

Each biological neuron is connected to a large number of other neurons. Electrical signals travel along these connections. These signals arrive at the neurons along the connections called the "dendrites." These signals produce a physiochemical reaction in the main body of the neuron called the "soma," which may result in generation of an electrical charge. The electrical charge causes a signal to travel along the "axon" and to be transmitted to the neurons along the synaptic connections. Figure 2.1 schematically shows the main elements of a biological neuron.

Neural networks are composed of a number of interconnected artificial neurons. A vast majority of the artificial neurons used in neural networks are based on the model proposed by McCulloch and Pitts in the 1940s (McCulloch and Pitts, 1943). The McCulloch–Pitts binary artificial neuron is illustrated in Figure 2.2.

Shown on the left-hand side of Figure 2.2 are a number of incoming connections, transmitting the signals from the other artificial neurons. A numerical value, called the connection weight, is assigned to each connection to represent its effectiveness or its strength in transmitting the signals. The weight of the connection from node number j into node number i is w_{ij}, and the signal coming from the node number j is S_j. The incoming connections are modeling the dendrites in the biological neurons.

The artificial neuron itself represents the soma in its biological counterpart. The physiochemical reactions that take place within the soma and cause it to fire a signal are represented

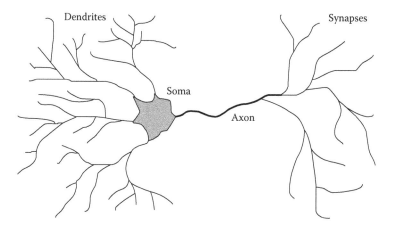

Figure 2.1 A simplified schematic representation of a biological neuron. The electrical signals generated by the other neurons arrive via the dendritic connections in the soma where a physiochemical reaction takes place and the generated signal travels along the axon and synaptic connection to other neurons.

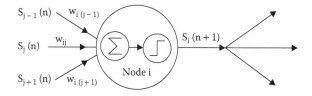

Figure 2.2 McCulloch–Pitts model of a binary artificial neuron with incoming weighted connections representing the dendrites and outgoing connections representing the synapses. The activation function is a step function and the activation of the artificial neuron is a binary number.

by two simple operations shown in the two circles. The first operation is to calculate the weighted sum of all the incoming signals, each weighted by the connection weight on which it is traveling.

$$z_i(n+1) = \sum_i w_{ij} S_j(n) - \theta_i \tag{2.1}$$

In this equation θ_i is the bias of the neuron. In the software implementation of the neural networks, the artificial neurons operate in discrete time intervals. In Equation 2.1, n denotes the discrete time station $t_n = n\Delta t$, where Δt is the time step. In reality, the operation of the artificial neurons is not affected by time, and n can represent discrete steps.

The second operation within the artificial neuron consists of passing the weighted sum through an *activation function*, f(x). The result of this operation is called the *activation* of the neuron, and it is denoted by $S_i(n + 1)$. Activation functions are usually bounded functions varying between 0 and 1, and they provide the main source of nonlinearity in neural networks.

In the McCulloch–Pitts binary artificial neuron, the activation function is a step function that determines whether the artificial neuron generates a signal or not.

$$S_i(n+1) = f\big[z_i(n+1)\big] = \begin{cases} 1 & \text{if } z_i(n+1) > 0 \\ 0 & \text{if } z_i(n+1) \leq 0 \end{cases} \tag{2.2}$$

The step function determines the binary activation of the artificial neuron. If the activation is equal to 1, then it is the signal that goes out of the artificial neuron along the outgoing connections, shown on the right-hand side of Figure 2.2, and travels to the other artificial neurons.

Another variation of the McCulloch–Pitts artificial neuron that is widely used is a real-valued neuron shown in Figure 2.3. The activations of these artificial neurons are real-valued number in the range of [0, 1] or [–1, 1]. The most commonly used activation function is the sigmoid function given in the following equation:

$$S_i(n+1) = f\big[z_i(n+1)\big] = \frac{1}{1 + e^{-\lambda z_i(n+1)}} \tag{2.3}$$

The sigmoid function is a smooth version of the binary step function and similar to the step function it varies between 0 and 1. However, the transition is more gradual, and it has a real non-zero value for all the possible values of its argument. The sigmoid function and the step functions are illustrated in Figure 2.4. Also shown in this figure are several sigmoid

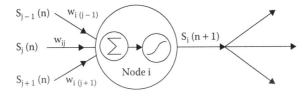

Figure 2.3 A real- valued artificial neuron with incoming weighted connections representing the dendrites and outgoing connections representing the synapses. The activation function is a real-valued bounded function and the activation of the artificial neuron is a real number.

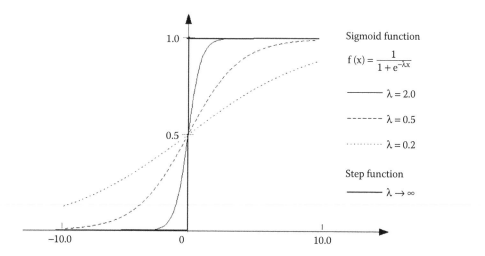

Figure 2.4 Sigmoid function used as the activation function of artificial neuron for several values of the parameter λ that controls the sharpness of the transition from 0 to 1, and the limiting case of the step functions.

functions with different values of λ. The value of λ determines the sharpness of the transition from 0 to 1. As the value of λ increases, the sigmoid function approaches the step function, and in the limit for λ → ∞ it coincides with the step function.

Another common choice for the activation function is the hyperbolic tangent function that is bounded function varying between –1 and 1.

$$f(x) = \tanh(ax) \tag{2.4}$$

The hyperbolic tangent function is shown in Figure 2.5 for several values of the parameter a that controls the slope of the function in transition from –1 to 1. The limiting shape of the function, as the parameter a approaches infinity, is the step function.

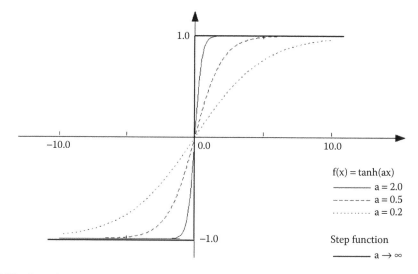

Figure 2.5 The hyperbolic tangent function used as the activation function of artificial neuron for several values of the parameter "a" that controls the sharpness of the transition from −1 to 1 and the limiting case of the step functions.

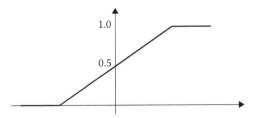

Figure 2.6 Bounded linear activation function.

Another function that is sometimes used as the activation function is the bounded linear function. The linear function bounded between 0 and 1 is shown in Figure 2.6. The linear function bounded between –1 and 1 can also be used as the activation function.

The activation function is the main source of the nonlinearity in the neural networks that are responsible for most of their complex behavior. Without the nonlinearity of the activation functions or with a linear activation function, the operation of the artificial neural network will reduce to a simple inner product of two vectors represented by the weighted sum.

2.3 GENERAL REMARKS ON NEURAL NETWORKS

Neural networks are constructed by connecting a number of artificial neurons, with each connection having its own strength in signal transmission. As we have seen in the previous section, each artificial neuron is only capable of performing very simple functions. A basic principle that also applies to neural networks is that the interaction between a large number of simple elements can produce complex behavior. Examples of such complex systems are brains and the nervous systems in humans and animals whose elements are the neurons; the immune system, whose elements are different types of white blood cells; the economy, the elements of which are the individuals, agencies, and firms; and the ecosystems whose elements are the species of plants and animals. In a similar vein, the interaction between a large number of artificial neurons in neural networks gives them the capability of exhibiting complex behavior, although in a much smaller scale than other complex systems in nature mentioned earlier. Complex systems have been discussed extensively in another volume (Ghaboussi and Insana, 2017).

It is important at the outset to make comparisons between artificial neural networks and brains in humans and animals. To start with, the operation of neurons and brains are not fully understood. What is known about the neurons seems to indicate that they are very different than the artificial neurons described in the previous section. In terms of scale, the human brain is estimated to have the order of 10^{11} to 10^{12} neurons and 10^{15} connections, whereas the largest of the current generation of neural networks in use have at most several thousand artificial neurons. In Chapter 4, we will discuss an application from the author's own research that used the neural network with about 3,000 artificial neurons. Of course, much larger neural networks are used in recognition problems. However, most neural networks used in engineering problems are much smaller.

Another important distinction is that the brain may have more than 100 different types of specialized neurons, whereas most neural networks use only one type of artificial neuron. The operations of the neurons are also quite different than the operation of artificial neurons. The electrical signals that the neurons generate consist of a series of spikes that can carry far more information than the single-value signals generated by the artificial neurons.

These differences between the biological and artificial neurons may explain the far more effective ways through which the neurons are used in the nature. The nervous systems of some animals contain fewer neurons than the largest neural network. For example, the nervous system of some worms is known to contain no more than 1,000 neurons, whereas in Chapter 4 we will discuss a neural network with about 3,000 neurons. However, even a lowly worn is a highly complex animal and its nervous system has far more capability than the larger artificial neural networks.

In spite of the major differences between neural networks and their biological counterparts, there are enough similarities to justify the frequently used statement that the neural networks are inspired by the brains and nervous systems in humans and animals. The primary similarity is that they are both "massively parallel" systems. The term massively parallel is used to describe the high level of connectivity between the neurons. Each neuron in the human brain has connections to 1,000 to 10,000 other neurons that makes every neuron within at most four neurons away from every other neuron. Neural networks are also massively parallel systems although in much smaller scale, and it is this high level of interconnectivity that allows the artificial neurons to interact and the neural network to exhibit complex behavior.

One very important aspect of the complex behavior in neural networks is their capability of learning. Learning in neural networks is classified as "supervised learning" and "unsupervised learning." These terms will be described in more detail later in this chapter. We will also see that the learning property of neural networks makes them most useful in engineering applications, especially in applications in computational mechanics.

2.3.1 Connecting artificial neurons in a neural network

Before discussing special types of neural networks and how their artificial neurons are connected, we will take a more fundamental look at this problem. The basic question can be posed in the following way. Is a special system of connecting the artificial neurons responsible for the important attributes of neural networks, such as learning? To find an answer to this question, again we go back to the brains and the nervous system in humans and animals. It was stated earlier that the human brain is thought to contain 10^{11} to 10^{12} neurons and 10^{15} connections. Are these neurons connected in a special and predetermined way or are the connections more random in nature? If the neurons were connected in a special way, the information about which neuron should be connected to which other neurons must be contained in the human DNA. The human DNA does not have the capacity to store that much information. This seems to indicate that the functions of the human brains are to a great extent independent of precisely how the neurons are connected and the connectivity may be more random.

One implication of the observation on the random nature of the connections in human brains is that the learning property of neural networks may not be dependent on the precise system connectivity of the artificial neurons. In other words, if we randomly connect a group of neural networks, the resulting neural network will also have the capability of learning. A randomly connected neural network is shown in Figure 2.7. In this neural network, each artificial neuron is connected to a randomly chosen number of other neurons.

These types of randomly connected neural networks have not been sufficiently developed for engineering applications. However, they hold promise, and in the future they may be developed into a powerful class of neural networks. Current equivalents of these neural networks are the fully connected networks. An example of fully connected neural network is shown in Figure 2.8. These fully connected neural networks have certain useful properties, and they will be discussed later in this chapter.

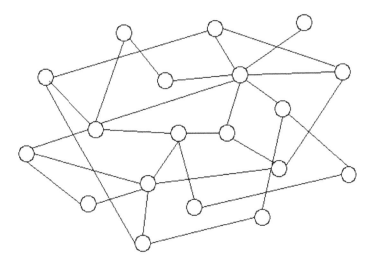

Figure 2.7 A randomly connected neural networks. Each neuron is connected to a random selection of other neurons.

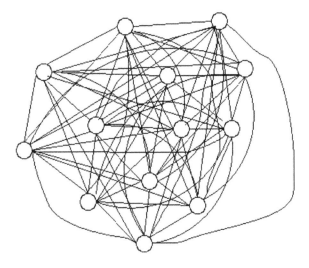

Figure 2.8 A fully connected neural network. Each neuron is connected to all the other neurons.

The fully connected neural network shown in Figure 2.8 is also referred to as the Hopfield net (Hopfield, 1982). Among the neural networks, Hopfield net is probably the closest to the brains in its operation. Similar to the brain, the Hopfield nets are dynamical systems, in the sense that a pattern of activations of the artificial neurons initiates the self-sustaining dynamic operation of networks. Each artificial neuron receives signals, performs the two tasks of weighted sum and activation function, and issues a signal that travels to all the other artificial neurons. This process can continue forever. However, Hopfield nets do settle into a stable state. Hopfield showed that these stable states correspond to the minima of a function that he called the *energy function* by observing the similarity with the mechanical systems whose equilibrium states correspond to the minimum of the potential energy.

2.4 PERCEPTRONS

Perceprtron consists of artificial neurons that are arranged in a layer called the output layer. Below the layer of artificial neurons is the layer of input nodes called the input layer. Each input node is connected to all the artificial neurons in the output layer. The single-layer perceptron is the simplest of the multilayer feedforward (MLF) neural networks that are the subject of the next section. Perceptrons were introduced in the early 1960s, and their limitations as neural networks were recognized a few years later. We will see later that due to these limitations, perceptrons are not useful neural networks. However, a knowledge of perceptrons will help in understanding the MLF neural networks.

A typical perceptron is shown in Figure 2.9. It has six input nodes and eight artificial neurons in its output layer. It is a fully connected neural network in the sense that each input node is connected to all the artificial neurons in the output layer. The input nodes receive the input values xj and transmit them to the artificial neurons. The efficiency in transmission of signals is defined through the numerical values of the connection weights. The artificial neurons perform the two basic operations of weighted sum and activation.

$$z_i = \sum_j w_{ij}x_j \qquad (2.5)$$

$$y_i = f(z_i) \qquad (2.6)$$

The activations of the artificial neurons, y_j, are the output of the perceptron. In perceptrons, the signals travel in one direction, from the input layer to the output layer. In general, one can think of a perceptron as a method of relating an input vector $\mathbf{X} = [\, x_1, \cdots, x_N]$ to an output vector $\mathbf{Y} = [y_1, \cdots, y_M]$, where N is the number input nodes and M is the number of artificial neurons in the output layer. The perceptron represents the following generic function:

$$\mathbf{Y} = F(\mathbf{X}) \qquad (2.7)$$

When we consider how the output of the perceptron, which is the activations of the artificial neurons, is calculated from the input values, it becomes obvious that the output is influenced by the values of the connection weights. There exists a specific set of connection weights w_{ij}

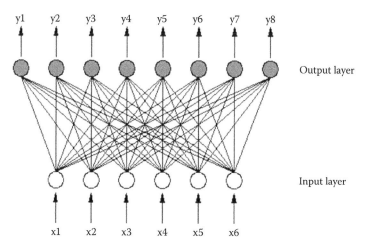

Figure 2.9 A single layer perceptron consisting of an input and an output layer. Input nodes are fully connected to all the nodes in the output layer. The artificial neurons are all in the output layer.

that will approximate the function in Equation 2.7. However, the values of those connection weights are unknown. The task of training a perceptron is to determine the specific set of connection weights that for any input vector it will give an output vector that approximates Equation 2.7. The trained perceptron is said to have learned the information content of the training dataset, to be described later. The information learned by the perceptron is stored in its connection weights. We will describe training of perceptrons as a special case of training multilayer neural networks in Section 2.6.

2.4.1 Linearly separable classification problems

A perceptron can be considered to be an assemblage of a number of perceptrons, each with one output node and the same number of input nodes as in the original perceptron. This is true as there is no connection between the output nodes. This point is illustrated in Figure 2.10. Each of the component perceptrons with the single output node is a classification perceptron, especially if the nodes are binary. A binary classification perceptron classifies its input vectors into two group. The first group comprises all the input vectors that cause an output value of one, and the second group comprises those vectors that cause a zero output. Similar classification is also possible when the activation function is real valued. In studying the properties of perceptrons, it suffices that we study a typical single output node classification perceptron.

In the mid-1960s, it was shown that the single-layer perceptrons are only capable of learning to classify linearly separable classification problems. When the input vectors can be separated into two classes by a hyperplane, then the classification is linearly separable. A linearly separable classification problem for a number of two-dimensional discrete points is shown in Figure 2.11. In a two-dimensional vector space, the hyperplane becomes a line.

Examples of linearly separable classification problems in two-dimensional binary sets $[x_1, x_2]$ are OR and AND. The function OR is defined as {output = 1 if $x_1 = 1$ OR $x_2 = 1$}, the function AND is defined as {output = 1 if $x_1 = 1$ AND $x_2 = 1$}. The classification problem OR and its four input–output pairs are shown Figure 2.12. It is clear that a line can separate the point with the output of 0 from the other three points with the output of 1.

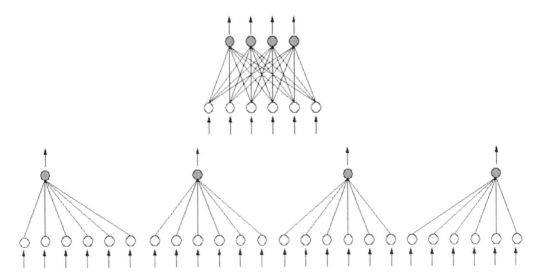

Figure 2.10 Illustration of the fact that the perceptron shown at the top can be considered as the assemblage of the four perceptrons shown below, each with a single-output node.

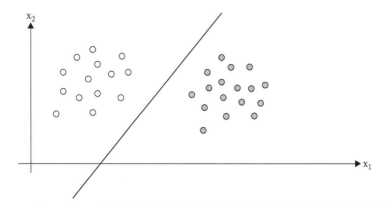

Figure 2.11 Two clusters of discrete points in two dimensions that form a linearly separable classification problem because a hyperplane (a line in 2D) can separate the two clusters.

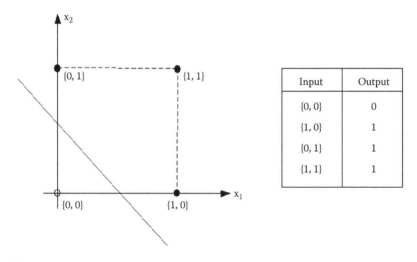

Input	Output
$\{0, 0\}$	0
$\{1, 0\}$	1
$\{0, 1\}$	1
$\{1, 1\}$	1

Figure 2.12 The binary linearly separable classification problem of OR. The two groups of point can be separated by a line.

Similarly, the classification for AND can be achieved in Figure 2.12 by a line that separates the point $\{1, 1\}$ from the other three points. In addition to OR and AND, there are four more linearly separable classification problems possible in the two-dimensional binary case.

2.4.2 Nonlinearly separable classification problems

The fact that the perceptrons can only learn to correctly classify linearly separable problems is a severe limitation as a vast majority, if not almost all, of the practical classification problems are nonlinearly separable. The groups in the classification problem are separated by a more complex hypersurface in the input vector space. A two-dimensional nonlinearly separable problem is illustrated in Figure 2.13. The two groups of discrete points cannot be

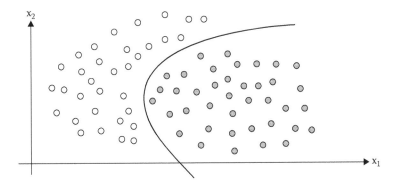

Figure 2.13 Two clusters of discrete points in two dimensions that form a nonlinearly separable classification problem because a general nonlinear surface (a curve in 2D) is needed to separate the two clusters.

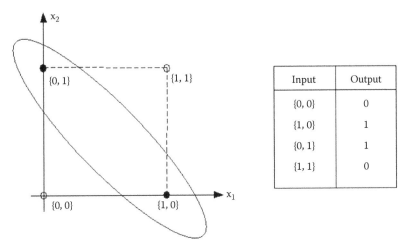

Input	Output
{0, 0}	0
{1, 0}	1
{0, 1}	1
{1, 1}	0

Figure 2.14 The binary nonlinearly separable classification problem of XOR. The two groups of point cannot be separated by a line.

separated by a line (hyper-plane in 2D). A nonplanar surface is required to separate these two groups and to classify them.

In the case of two-dimensional binary classification, XOR (exclusive OR) is a nonlinearly separable problem. The function XOR is defined as {output = 1 if $x_1 = 1$ and $x_2 = 0$ or $x_1 = 0$ and $x_2 = 1$}. The function and its input–output pairs are shown in Figure 2.14.

In addition to XOR, there is one more nonlinearly separable classification possible in the two-dimensional binary problems. The ratio of nonlinearly separable to linearly separable classification problems in the two-dimensional binary case is 2 to 6. However, this number is somewhat misleading. That ratio increases very fast as the dimension of the problem increases. In practical problems in which the dimension of the input vector space may be quite large, the number of linearly separable problems is negligibly small compared with the number of nonlinearly separable problems. Almost all the practical problems are of the nonlinearly separable type.

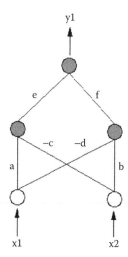

Figure 2.15 A two-layer neural network with the appropriate connection weight for correctly classifying the XOR problem.

When the limitations of the perceptrons became known in the mid-1960s, it was also shown that neural networks with more than one layer of artificial neurons are needed to correctly learn the nonlinearly separable problems. In the case of the XOR problem, the two-layer binary neural network shown in Figure 2.15 can correctly classify the problem. The activation functions in this neural network are step functions and connection weights are a, b, c, d, e, f > 0, c ≥ a and d ≥ b. It can easily be verified that first layer of neurons will produce the following activations [0, 0], [1, 0], [0, 1], [0, 0] for the input vectors [0, 0], [1, 0], [0, 1], [1, 1], respectively, and these will produce the correct activations of 0, 1, 1, 0 at the output node.

2.5 MULTILAYER FEEDFORWARD NEURAL NETWORKS

MLF neural networks are probably the most widely used neural networks. With a few exceptions, the vast majority of the neural network applications in engineering mechanics described in this book use the MLF neural networks. Unlike the randomly connected or the fully connected Hopfield nets, the MLF neural networks are not dynamical systems and consequently; they least resemble the nervous system in humans and animals.

Similar to perceptrons discussed in the previous section, the artificial neurons in the MLF are arranged in a number of layers. The first layer is the input layer, and the last layer is the output layer. The layers between the input and the output layers are referred to as the *hidden layers*. The order of the layers and the direction of the propagation of the signals is from the input layer, through the hidden layers to the output layer. In the fully connected version, each node is connected to all the nodes in the next layer. Figure 2.16 shows a typical MLF neural network.

Again, similar to the perceptrons, the nodes in the input layer are not quite artificial neurons. They only receive the input values and transmit them to the artificial neurons in the first layer that is the first hidden layer.

There are two different ways of counting the layers in MLF neural networks. Some prefer to count all the layers including the input layer. For example, the MLF neural network shown in Figure 2.16 is referred to as a four-layer neural network, whereas others prefer to only count the layers with artificial neurons and to exclude the input layer. In this case, the

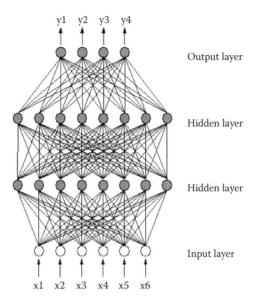

Figure 2.16 A multilayer feedforward neural network consisting of an input layer, two hidden layers, and an output layer.

MLF neural network shown in Figure 2.16 is referred to as a three-layer neural network. In this book, we will adopt the later convention and not count the input layer in describing the neural network.

The type of fully connected neural network shown in Figure 2.16 is the most commonly used. However, in Chapter 3 we will see that other patterns of connections are also possible. Less than the fully connected neural networks arise when pruning techniques are used to remove the connections that are deemed unnecessary according to some criterion. Other patterns of connectivity can be the result of some special structure in the training data. We will describe examples of the latter case, known as nested adaptive neural networks, in Chapter 3.

The nodes in the MLF neural networks are the typical artificial neurons that were described earlier. The activation of the nodes is determined from an activation function and a weighted sum operation.

$$z_i^k = \sum_j w_{ij}^k \, S_j^{k-1} - \theta_i \tag{2.8}$$

$$S_i^k = f\left(z_i^k\right) \tag{2.9}$$

The superscript k is used to designate the layer number that varies from 0 for the input layer to n for the output layer. In Equation 2.8 θ_i is the bias, w_{ij}^k is the weight of the connection from node j in layer k–1 coming into node i in layer k, and S_i^k is the activation of node number i in layer number k. The input vector can be considered as the activations of the input nodes and the activation of the output nodes are the output of the neural network.

$$S_i^0 = x_i \tag{2.10}$$

$$y_i = S_i^n \tag{2.11}$$

The activation function for the nodes is a bounded function varying between 0 and 1 or between –1 and 1. In binary neural networks, the activation function is a step function. In the real-valued neural networks, the activation function is either a sigmoid $f(x) = 1/(1 + e^{-\lambda x})$, or hyperbolic tangent f(x) = tanh(ax), or a bounded linear function as shown in Figures 2.4 through 2.6.

2.5.1 A notation for multilayer feedforward neural networks

Describing neural networks in words so that they can be duplicated may be too cumbersome. A compact notation will be especially helpful in formulating problems in which the neural network may be a component of the formulated problem. In this book, we will use a notation that was first introduced in Ghaboussi et al. (1998) and has been used in a number of publication by the author, his coworkers, and other researchers.

The notation describing a neural network should define the input and the output of the neural network. It is known that many neural networks with different architectures can be trained to learn the information in the training data with the same input and output vectors to within a reasonable level of accuracy. Therefore, the architecture of the neural network is a relevant information if the results are to be duplicated. The information about the architecture of the neural network should also be included in the notation describing the neural network.

The notation that will be used in this book has the following general format:

$$\mathbf{Y} = \mathbf{YNN}\left[\mathbf{X} : \text{network architecture}\right] \tag{2.12}$$

Or simply

$$\mathbf{Y} = \mathbf{NN}\left[\mathbf{X} : \text{network architecture}\right] \tag{2.13}$$

It is similar to the notation describing a mathematical function. The difference being that a neural network is approximating the function. The notation indicates that the vector $\mathbf{Y}$ is the output of a neural network. The first argument field in the parenthesis describes the input vector $\mathbf{X}$ and the second argument field describes the architecture of the neural network.

The neural network shown in Figure 2.16 is described by the following equation in the new notation:

$$\mathbf{Y} = \mathbf{NN}[\mathbf{X} : 6 \mid 8 \mid 8 \mid 4] \tag{2.14}$$

The argument field describing the network architecture indicates that it is a three-layer neural network: Input layer has six nodes, each of the two hidden layers has eight nodes, and the output layer has four nodes. In some cases, it may be necessary to include the terms of the input and output vectors individually. The neural network in Figure 2.16 can then be described by the following equation:

$$\{y_1, \ldots, y_4\} = \mathbf{NN}\left[\{x_1, \ldots, x_6\} : 6 \mid 8 \mid 8 \mid 4\right] \tag{2.15}$$

In some cases, the notation is used to describe a generic neural network and the specific architecture is not known or need not be specified. This situation will be encountered frequently in this book when we describe a particular formulation that uses neural networks, without specifying a specific neural network. In such cases, the argument field for the network architecture will be left blank, as in the following equation:

$$Y = NN[X: \quad]$$ (2.16)

This equation indicates that vector Y is the output of some neural network whose input vector is X without specifying the specifics of the neural network architecture.

Later we will describe a method of adaptive architecture determination. In these types of neural networks, the history of the adaptive evolution of the network architecture during the training is also relevant information for duplicating the results of the neural network. We will also include some information on the history of the adaptive evolution of the network architecture in the second argument field within the parenthesis. As we will see later, there are many other parameters in the training of the neural networks. However, it is impractical to include all the details of the training in the second argument field. In fact, most of the parameters are not that essential in the performance of the neural network.

2.6 TRAINING OF MULTILAYER FEEDFORWARD NEURAL NETWORKS

The response (output) of an MLF neural network to any given stimuli (input) obviously will depend on the connection weights. This can clearly be seen in Equations 2.8 through 2.11. The choice of the activation function also has an influence on the stimulus–response behavior of neural network. However, the activation function is a fixed part of the neural network, and it does not change during the training of the neural network. The training of a neural network essentially means the adaptation of the connection weights.

2.6.1 Supervised learning

The training of the MLF neural networks is termed "supervised learning" as the neural network learns from the dataset of input/output pairs. The knowledge to be learned and acquired by the neural network is contained in the set of input/output pairs that constitutes the training dataset as shown in the following equation:

$$[Y_1, X_1], \cdots, [Y_k, X_k]$$ (2.17)

Before the training starts, the connection weights are assigned random values. During the training, the connection weights of the neural network are changed so that for any input from the training dataset, the output of the neural network matches the corresponding output from the dataset as closely as possible. At the successful completion of the training, the connection weights of the network have captured and stored the underlying knowledge present in the training dataset. When presented with an input pattern, the forward pass through the neural network results in an output pattern, which is the result of the generalization and synthesis of what it has learned and stored in its connection weights.

We can think of supervised learning as an optimization problem to minimize the output error of the neural network for all the cases in the training dataset. Any optimization method can be applied to this problem. Supervised learning is an iterative process that gradually changes the connection weights.

2.6.2 Backpropagation

Backpropagation is a form of gradient decent optimization method in minimizing the output error. The output error for the input X_p from the training dataset is defined in the following equation in terms of the difference between the target output Y_p from the training dataset and the output of the neural network $\bar{Y}_p$.

$$e_p = \frac{1}{2}\left\|Y_p - \bar{Y}_p\right\|_2^2 = \frac{1}{2}\sum_{i=1}^{M}\left(y_{pi} - \bar{y}_{pi}\right)^2 \tag{2.18}$$

The total error E is the sum of the errors for all the input/output pairs in the training dataset:

$$E = \sum_p e_p \tag{2.19}$$

Obviously, the total error is a function of connection weights of the neural network:

$$E = E\left(w_{ij}\right) \tag{2.20}$$

The essence of the training of a neural network is to determine a set of connection weights that minimizes the total error E. The rules used to update the connection weights are called the learning rule.

As we mentioned earlier, any optimization method can be used to determine the optimal connection weights. The most commonly used method is an iterative method of updating the connection weights based on a simple variation of the gradient descent method.

$$\Delta w_{ij} = -\eta\,\frac{\partial E\left(w_{ij}\right)}{\partial w_{ij}} \tag{2.21}$$

In this equation η is the *learning rate*. It is usually a small number between 0 and 1. Learning rate is an important parameter that governs the rate of convergence of the gradient-based algorithm. We will revisit the learning rate and discuss it in more detail later.

Derivative of the total error with respect to the connection weights is the sum of the derivatives of the input/output pairs in the training dataset:

$$\frac{\partial E}{\partial w_{ij}} = \sum_p \frac{\partial e_p}{\partial w_{ij}} \tag{2.22}$$

As the derivative of the total error is the sum of derivatives of the errors of each pattern, it suffices that we determine the derivative of one pattern with respect to the connection weights.

At this point, we introduce a superscript to designate the layer number. The connection weight w_{ij}^k connects the node i in layer k–1 to node j in layer k. This connection weight obviously affects the activation S_j^k of the node j in layer k, leading to the following expression:

$$\frac{\partial e_p}{\partial w_{ij}^k} = \frac{\partial e_p}{\partial S_j^k}\,\frac{\partial S_j^k}{\partial z_j^k}\,\frac{\partial z_j^k}{\partial w_{ij}^k} \tag{2.23}$$

The last term on the right-hand side of Equation 2.23 is determined by using Equation 2.8 for the weighted sum of the incoming signals.

$$\frac{\partial z_j^k}{\partial w_{ij}^k} = \frac{\partial}{\partial w_{ij}^k}\left(\sum_m w_{mj}^k S_m^{k-1} - \theta_m\right) = \frac{\partial}{\partial w_{ij}^k}\left(w_{ij}^k S_i^{k-1}\right) = S_i^{k-1} \tag{2.24}$$

For the special case of k = 1, the layer above the input layer, we have $S_i^{k-1} = S_i^0 = x_i$ – the incoming activations are the input to the neural network from the input layer.

The second term in Equation 2.23 is simply the derivative of the activation function.

$$\frac{\partial S_j^k}{\partial z_j^k} = \frac{\partial}{\partial z_j^k} f\left(z_j^k\right) = f'\left(z_j^k\right) \tag{2.25}$$

For the special case of sigmoid activation function, we have the following expression:

$$\frac{\partial S_j^k}{\partial z_j^k} = f\left(z_j^k\right)\left[1 - f\left(z_j^k\right)\right] \tag{2.26}$$

The first term in Equation 2.23 has two forms, depending on the location of the layer k in the neural network. If k = n, it is the output layer and the activation of node j is the output of the neural network $S_j^k = \bar{y}_j$, leading to the following expression, by using the definition of the output error from Equation 2.18.

$$\frac{\partial e_p}{\partial S_j^k} = \frac{\partial e_p}{\partial \bar{y}_j} = \frac{\partial}{\partial \bar{y}_j}\left[\frac{1}{2}\sum_{i=1}^{M}\left(y_{pi} - \bar{y}_{pi}\right)^2\right] = \left(\bar{y}_{pj} - y_{pj}\right) \tag{2.27}$$

If the layer k is a hidden layer, then the first term in Equation 2.23 is somewhat more complicated. We observe that the activation S_j^k affects the weighted sum of all the nodes in the layer k + 1. This observation leads to the following derivation:

$$\frac{\partial e_p}{\partial S_j^k} = \frac{\partial e_p\left(z_1^{k+1}, \ldots, z_m^{k+1}\right)}{\partial S_j^k} = \sum_{r=1}^{m}\frac{\partial e_p}{\partial z_r^{k+1}}\frac{\partial z_r^{k+1}}{\partial S_j^k}$$

$$= \sum_{r=1}^{m}\frac{\partial e_p}{\partial S_r^{k+1}}\frac{\partial S_r^{k+1}}{\partial z_r^{k+1}}\frac{\partial z_r^{k+1}}{\partial S_j^k} \tag{2.28}$$

$$= \sum_{r=1}^{m}\frac{\partial e_p}{\partial S_r^{k+1}} f'\left(z_r^{k+1}\right) w_{jr}^{k+1}$$

$$= \sum_{r=1}^{m}\xi_r^{k+1} w_{jr}^{k+1}$$

$$\xi_j^k = \frac{\partial e_p}{\partial S_j^k} f'\left(z_j^k\right) \qquad (2.29)$$

Combining Equations 2.23 through 2.29, we arrive at the following equation:

$$\begin{cases} \dfrac{\partial e_p}{\partial w_{ij}^k} = \xi_j^k\ S_i^{k-1} \\[4mm] \xi_j^k = \dfrac{\partial e_p}{\partial S_j^k}\ \dfrac{\partial S_j^k}{\partial z_j^k} = \begin{cases} \left(\overline{y}_{pj} - y_{pj}\right)f'\left(z_j^k\right) & \text{for } k = n;\ \text{output layer} \\[4mm] \left(\displaystyle\sum_{r=1}^{m} \xi_r^{k+1} w_{jr}^{k+1}\right)f'\left(z_j^k\right) & \text{for } k = 1, \cdots, n-1;\ \text{hidden layers} \end{cases} \end{cases} \qquad (2.30)$$

In the case of perceptrons, as there is only one layer and that is the output layer, the first part of the equation above for $k = n$ applies, and there is no backpropagation.

In the case of MLF neural network, we do have backpropagation that starts from the output layer and works down to the layer $k = 1$ that is above the input layer. We can see from Equation 2.30 that we start at the output layer $k = n$ and compute all ξ_j^n. Then we move to the lower part of Equation 2.30 for hidden layer $k = n-1$ and determine all ξ_j^{n-1}. This process is continued to the bottom layer for $k = 1$. Now we can see why this training method is called backpropagation, starting from the output layer back to layer one. Updating of the connection weights takes place with the following equation:

$$\begin{aligned} \Delta w_{ij} &= -\eta\ \frac{\partial E\left(w_{ij}\right)}{\partial w_{ij}} \\[3mm] &= -\eta\ \sum_p \xi_j^k\ S_i^{k-1} \end{aligned} \qquad (2.31)$$

where η is the learning rate. The negative sign is needed because the changes need to occur downward along the gradient to minimize the output error.

2.6.3 Discussion of backpropagation

In the previous section, we described the backpropagation method of training MLF neural networks and pointed out that it is based on the gradient decent method. There are several important issues that need to be discussed. First important issue involves updating of the connection weights.

2.6.3.1 Updating of connection weights

The training data consist of a number of input/output pairs. For each input/output pair, we determine the changes to the connection weights. One important issue concerns the updating of the connection weights.

Updating the connection weights for all training input/output dataset is called an epoch. We can think of updating the connection weights as an iterative process and each iteration is called an *epoch*.

Connection weights can either be updated for each training dataset or they are normally updated for *batches* of training cases. For a small number of training cases, the batch may include all the training cases. Changes in connection weights are computed for all the training cases in the batch, and then they are updated.

When there are many training cases, a smaller batch of training cases are selected. Often the batches are selected randomly for each epoch, or they are used for several epochs, before randomly selecting a new batch of training cases.

Next, we need to discuss the convergence of the iterative process of training backpropagation. There are two important issues that have been extensively studied by researchers in this field. There is often concern that the gradient decent method that backpropagation is based on may get stuck in a local minimum. The other issue concerns the rate of convergence. Many variations of the basic backpropagation method have been proposed. We will briefly discuss few of these methods.

One method is *adaptive learning rate*. In Equation 2.31, the change in connection weights is the negative of the gradient of the error multiplied by the learning rate η. In most cases, the learning rate is not constant; it changes during the training process. The changes in the output error are monitored and the learning rate is modified accordingly. Smaller learning rate reduces the chance of the training process getting stuck in a local minimum. Changing the learning rate may also avoid overtraining, which will be discussed in Section 2.9.

Another method of avoiding local minima or getting out of local minima during the training is *Backpropagation with momentum*. The connection weights are updated on the basis of the current and previous epoch, as shown in the following equation:

$$\Delta w_{ij}(n) = (1 - \alpha)\left(-\eta \sum_p \xi_j^k \, s_i^{k-1}\right) + \alpha \, \Delta w_{ij}(n-1) \tag{2.32}$$

In this equation, n is the epoch number and $0 < \alpha < 1$ is the inertia constant. The addition of momentum is often effective and is commonly used in fixing the problem of oscillating gradient descent.

There are also probabilistic or stochastic neural networks that introduce some randomness into the backpropagation method that can help avoid the local minima. In some cases, the training dataset is stochastic, and it requires stochastic neural network to represent the problem. Usually, some form of randomness is introduced in the activation function. The simplest form is the *Gaussian Machine* that adds a random variable to the computed activation of the neurons. There is also *Boltzmann Machine* that uses a probabilistic form of the sigmoid activation function. We will present a new form of stochastic neural network in Chapter 4, Section 4.8 that was used in an inverse problem in earthquake engineering.

A number of other variations of basic backpropagation method have been introduced to avoid the local minima and speed up the convergence, such *Quickprop, Delta-bar-delta, Rprop,* and *Steepest decent method*.

Another possibility for training the MLF that involves minimizing the output error is to use *Genetic algorithm* that will be described in Chapter 6.

2.6.4 Training and retraining

In this section, we will briefly discuss some important aspects of training of neural networks. In the previous chapter, we discussed some fundamental properties of soft computing methods, including neural networks. These properties were imprecision tolerance, nonuniversality, and functional nonuniqueness. We will demonstrate some aspects of these properties using a simple example. A simple neural is trained to learn the function y = sin(x). Of course, we know that this function is universal, meaning that it is valid for $-\infty<x<\infty$. We have also mentioned that neural networks cannot learn universal data. They can learn, with some imprecision tolerance the function y = sin(x) over a range of x. Outside that range, neural networks exhibit their nonlinear behavior. One neural network with one input and one output node and two hidden layers, each with three nodes, is trained over two ranges of the function and results are shown in Figure 2.17. On top of the figure, the two ranges are shown. Within each range of x, 100 points were randomly selected and used to train two neural network NN_1 and NN_2. The responses of the trained neural networks are shown at the bottom part of the figure. This example clearly demonstrates the nonuniversality in the responses of the trained neural networks. Unlike mathematical functions, neural networks can learn only the information contained in the data and that information cannot be universal. We can also see the imprecision in the response of the trained neural networks. Another important point to observe is that neural networks with the same architecture are able to learn different amounts of information, until they approach their capacity to store the information. We can see that this is almost the case in NN_2, on the right.

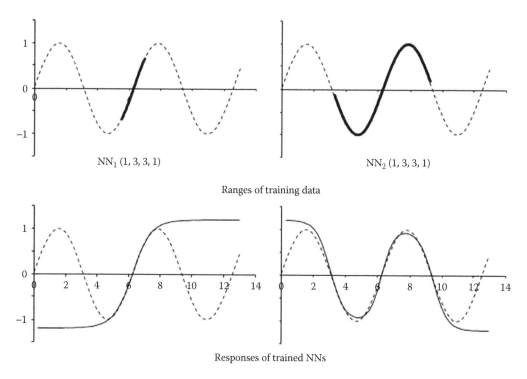

Figure 2.17 A simple neural network is trained to learn the y = sin(x) over two ranges of x, shown at the top, and the response of the trained neural networks are shown at the bottom.

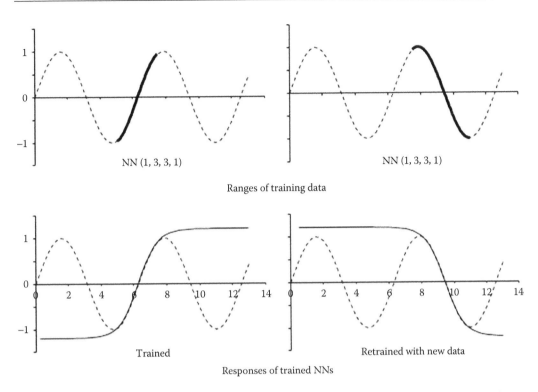

Figure 2.18 A simple neural network is trained with data shown on the left and continued training with new data shown on the right. NN unlearns the information in the old data and learns the information in the new data.

Neural networks cannot only learn information in the training data, they can also unlearn that information and learn information in some new training data. This is demonstrated in the simple neural with architecture NN (1, 3, 3, 1) trained to learn y = sin(x), shown in Figure 2.18. On the left the NN is trained with 100 randomly selected points in the range shown at the top and the response of the trained NN is shown at the bottom left. Next, the same NN is continued training with new data from the range shown at the top right. The response of the trained NN is shown at the bottom right. It is clear that the neural unlearned (or forgot) the old information and learned the new information.

Often, we encounter situations where a neural network is trained with some data, and later new information and training data become available. If the new data complements the information in the old data, the training can be continued with both sets of data. On the other hand, if the new information is basically different, then the training can be continued with only the new data and the NN replaces the information it has learned with the information in the new data. The second approach was used in the case shown in Figure 2.18.

The examples shown in this section were mainly intended to illustrate some important points. The data used in these examples was precise, without any noise and scatter. In most applications, we often deal with data that contain some information that is contaminated with noise and scatter. Important issues arise in those situations, and we have to deal with the issue of over training. These issues are discussed in Section 2.9.

2.7 HOW MANY HIDDEN LAYERS?

To start with, there are no rigorous general rules for determining the appropriate number of hidden layers. Like many aspects of neural networks, the number of hidden layers is problem dependent. The author's own experience, as well as a consensus among most of the users of neural networks, is that two hidden layers is sufficient for most problems. As the number of hidden layers increase beyond 2, the capacity of neural networks to learn increasingly more complex information increases.

The question of whether two or more hidden layers are needed depends to some extent on the nonlinearity and the complexity of the underlying information in the training data that the neural network is expected to learn. One hidden layer may be sufficient for some simple problems. If the problem can be solved and the neural network can be trained with one hidden layer, then it is preferable not to use two or more hidden layers for that problem. However, for many practical problems one hidden layer is not sufficient.

The neural network applications in engineering mechanics presented in this book all use two hidden layers and most of these problems cannot be solved with one hidden layer. This is because of the high degree of nonlinearity in the engineering mechanics problems presented in the later chapters in this book.

Of course, there are some exceptions to the rule of maximum of two hidden layers. As we will see later, there are some cases, like the replicator neural networks that may require three or more hidden layers. In some applications, we will see that a composite neural network may appear to have up to four hidden layers. However, these neural networks are composed of more than one neural network, and the constituent neural networks are trained separately.

2.8 ADAPTIVE NEURAL NETWORK ARCHITECTURE

Feedforward neural networks consist of several layers of artificial neurons or processing units: the input layer, the output layer, and several hidden layers. Two hidden layers are sufficient in most engineering applications. The number of processing units in the input and the output layers is dependent upon the formulation of the problem for neural network representation. The number of processing units in the hidden layers is a difficult part of the network architecture determination. The number of processing units in the hidden layers determines the capacity of a neural network, which in turn is related to the complexity of the underlying information in the training data. However, the degree of complexity of the problem cannot easily be quantified, and its relation to the size of the neural network is not well understood. It should also be pointed out that the representation problem does not have a unique solution; many different network architectures can produce similarly satisfactory results. Trial and error is one method of architecture determination. Adaptive determination of network architecture has played an important role in earlier neural network material modeling studies (Ghaboussi et al., 1990, 1991, 1994; Wu, 1991; Wu et al., 1992; Ghaboussi and Wu, 1998). The method for adaptive evolution of network architecture was first developed by (Wu, 1991) and has been further developed and refined (Joghataie et al., 1995).

In the adaptive method of architecture determination, which is shown schematically in Figure 2.19, the number of nodes in the hidden layers is adaptively determined during the training of the neural network. The training starts with an arbitrary, but small, number of neurons in the hidden layers. The learning rate is monitored during training, and as the network approaches its capacity (the rate of error reduction is reduced to a very small amount), new nodes are added to the hidden layers. The process of adding new nodes is

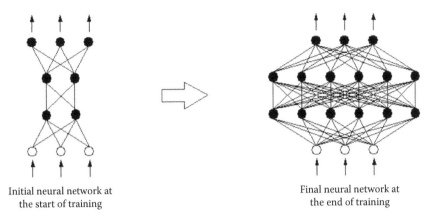

Initial neural network at
the start of training

Final neural network at
the end of training

Figure 2.19 The process of adaptive training of the neural network adds new nodes during the training.

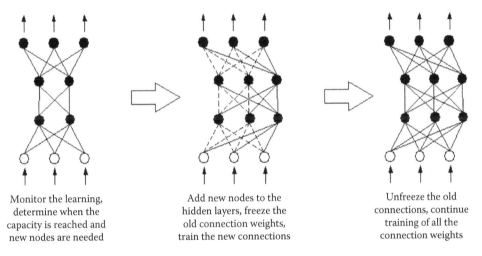

Monitor the learning,
determine when the
capacity is reached and
new nodes are needed

Add new nodes to the
hidden layers, freeze the
old connection weights,
train the new connections

Unfreeze the old
connections, continue
training of all the
connection weights

Figure 2.20 The procedure for the adaptive evolution of the neural network architecture during training of the neural network.

shown schematically in Figure 2.20. When new nodes are added, they generate new connections. The objective of the continued training, immediately after the addition of new nodes, is for the new connection weights to acquire the part of the knowledge base in the training data that was not acquired and stored in the existing connection weights. To achieve this, some training epochs are carried out with only the new connection weights being modified, whereas the old connection weights are frozen. This is followed by additional epochs of training where all the connection weights are allowed to change. These steps are repeated, and new nodes are periodically added to the hidden layers as needed. At the end of the training process, the appropriate network architecture has therefore been determined automatically. This method of adaptive network architecture determination has many advantages over fixed architecture neural networks. The most important advantage is that it is less likely to get stuck in local minima during training, which is a serious problem in fixed architecture neural networks. This assertion is based primarily on experience, observation, and intuition; it remains to be rigorously proven.

The adaptive architecture determination is an important part of the neural network training and, as such, it needs to be represented in the notation for neural networks that was introduced in Section 2.5.1. The adaptive architecture of neural network shown in Figure 2.19 is shown in the following equation:

$$Y = NN[X : 3 | 2 - 6 | 2 - 6 | 3] \tag{2.33}$$

This equation indicates that the neural network has three input and three output nodes and the hidden layers started with two nodes each and adaptively increased to six nodes at the end of training. When new data become available, the training can continue, and new nodes may be needed for the hidden layers.

In the next chapter, we will discuss nested adaptive architecture. We will see that the nested structure of the neural also needs to be presented in the notation, in addition to the adaptive node generation. This is important because the notation for neural network has to contain sufficient information to enable reproduction of the trained neural network.

2.9 OVERTRAINING OF NEURAL NETWORKS

Neural networks are sometimes trained with data generated by a numerical simulation such as finite-element analysis. These types of training data are often *clean*—they have no noise or scatter beyond the roundoff error. Neural networks can learn the information contained in clean data as closely as possible.

The situation with actual, real world data is different. Actual measured data often have noise and scatter as well as measurement error. This type of actual measured data contains information in addition to the noise, scatter, and measurement error. The objective is to train the neural network to learn the information contained in the data without also learning noise, scatter, and measurement error. This is the difference between training and overtraining. When a neural network is overtrained, it has learned the noise, scatter and the measurement error that can dominate the information in the training data. Overtraining can occur in two ways; the size of the neural network and the number of training epochs.

We will illustrate overtraining in two simple examples. In both examples, we have the training data shown in Figure 2.21. It is obvious that these data contain some information in the form of function y(x). It also has noise and scatter. In the first example, we start with training a small neural network with one input node, one output node, and two hidden

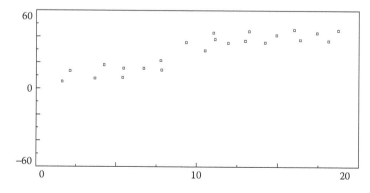

Figure 2.21 Training data with noise and scatter used to train a neural network with one input and one output and two hidden layers.

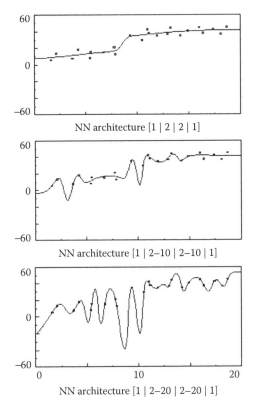

Figure 2.22 Three neural networks are trained with the data shown in Figure 2.21. The results from the basic neural network are shown at the top. The lower two show the results of training starting with the same simple neural network but adaptively increasing the number of nodes in the hidden layers.

layers, each with two nodes with the NN architecture [1 | 2 | 2 | 1]. We also use adaptive architecture, starting with the same simple neural network, and increase the hidden layer nodes from 2 to 10, in one case [1 | 2–10 | 2–10 | 1] and from 2 to 20 in the next case [1 | 2–20 | 2–20 | 1]. The three trained neural networks are tested, and the results are shown in Figure 2.22. We can see that the simple neural network with two nodes in the hidden layers shown at the top of Figure 2.22 has learned the main information in the training data. This means that this simple neural network has the capacity to learn that information but does not have the capacity to learn the noise and scatter in the data. We observe that the noise and scatter are far more complex than the main information in the training data. The larger neural networks appear to have the capacity to learn the noise and scatter in the data, and they obviously are doing so. At the bottom of Figure 2.22, the largest neural network with 20 nodes in the hidden layers seems to have mainly and precisely learned the noise and scatter in the data that hides the main information in the data that the simple neural network was able to learn.

The number of training epochs and the averaged error are shown in Table 2.1. Average error (AE) is defined in the following equation:

$$AE = \frac{1}{N} \sum_{j=1}^{N} \frac{1}{2} \left(t_j - o_j \right) \tag{2.34}$$

Where N is the number of data points, t is the target output, and o is the NN output.

Table 2.1 Number of training epochs and averaged errors for the cases
shown in Figure 2.22

NN architecture	Number of epochs	Averaged error (AE)
[1 | 2 | 2 | 1]	10,000	0.00256
[1 | 2–10 | 2–10 | 1]	90,000	0.001025
[1 | 2–20 | 2–20 | 1]	210,000	0.000036

In the next case, we start with a large neural network with the architecture [1 | 20 | 20 | 1] and train it with increasing number of epochs. Obviously, this neural network does have the capacity to learn the noise and scatter in the training data. However, this large neural network can learn the noise and scatter in data if we keep training it with higher and higher number of epochs. The results of training this large neural network with 500; 10,000; and 50,000 epochs are shown in Figure 2.23. Even though this is a very large neural network with a large capacity for information, after 500 epochs it seems to have learned the main information in the training data reasonably well as shown at the top of Figure 2.23. When we continue training it with increasing number of epochs, it starts learning the noise and scatter. We can see at the bottom section of Figure 2.23 that it has completely learned the noise and

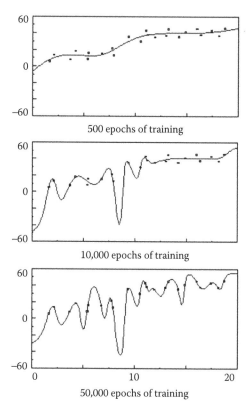

500 epochs of training

10,000 epochs of training

50,000 epochs of training

Figure 2.23 Performance of the neural network with the fixed architecture [1 | 20 | 20 | 1] at three different levels of training epochs.

Table 2.2 Fixed architecture [I | 20 | 20 | I] neural networks

Number of epochs	Averaged error (AE)
500	0.002736
5,000	0.002223
10,000	0.001055
20,000	0.000395
30,000	0.000174
40,000	0.000142
50,000	0.000000213

scatter in the data without the main information. The decrease in the averaged error with the increase in the training epochs is shown in Table 2.2. As the neural network learns more of the noise and scatter, the averaged error decreases.

We can see from Tables 2.1 and 2.2 that there is relationship between overtraining and the AE. As the neural network gets increasingly overtrained, AE error decreases. This provides a means of monitoring overtraining.

In the simple examples shown in Figures 2.22 and 2.23, the main information in the noisy training data is fairly simple that a small neural network can easily learn. In most real-world application, the main information in the noisy data that we want the neural network to learn is often fairly complex and requires large neural networks to learn that information. This opens up the possibility of overtraining and the neural network also to learn the noise and scatter, whereas it is learning the main information.

There are other means of preventing overtraining in large neural networks. In deep neural networks that are used for speech, text, and image recognition, overtraining may be prevented by pruning some artificial neurons randomly. One way of looking at this is that noise and scatter in the data are often of high frequency because of their random nature. On the other hand, the main information in the training data is of lower frequency. Because of the error tolerance of the neural networks, the main information that the neural network is learning is far less sensitive to random pruning of the few nodes than the noise and scatter in the data.

2.10 NEURAL NETWORKS AS DYNAMICAL SYSTEMS; HOPFIELD NETS

Hopfield nets were briefly mentioned in Section 2.3.1 and a fully connected Hopfield net was shown in Figure 2.8. We have seen that Hopfield nets consist of a number of neurons with binary activation function that are fully connected. The connectivity and connection weights must satisfy the following two conditions;

1. Neurons are not connected to themselves, $w_{ii} = 0$
2. Connections weights are symmetric, $w_{ij} = w_{ji}$.

We can think of connection weights as terms of a square symmetric matrix $\mathbf{W}$ with zero diagonal terms. Diagonal terms are 0 because they represent the neurons connection to themselves. Symmetry and 0 diagonal terms satisfy the two conditions mentioned above.

Activation functions are binary step functions with values –1 and 1. Some Hopfield nets also use binary activation step function 0 and 1. The activation functions have a threshold of a_i, as shown in the following equation:

$$S_i = \begin{cases} 1 & \text{if } \sum w_{ij}S_j \geq a_i \\ -1 & \text{if } \sum w_{ij}S_j < a_i \end{cases} \tag{2.35}$$

Hopfield nets are content addressable associative memories. They can store a number of binary pattern vectors that can be retrieved. They can be activated by assigning activations to the neurons. This initiates the Hopfield net to operate as a dynamical system. The network is updated at regular time steps. Update consists of neurons receiving signals from the other neurons and performing the two basic tasks of weighted sum of incoming signals and activation, and sending out its activation signal to other neurons. Updates can occur in two different ways.

Asynchronous update: Only one neuron is updated at each time step. This can be done either by randomly selecting the neurons to be updated, or by using a predefined order of selecting the neurons to be updated. In random selection of neuron for updating, it is important to take measures to avoid the possibility of frequent selection of the same neuron.

Synchronous update: All the neurons are updated simultaneously at regular time steps. We note that this is possible in computational simulation, but not feasible and likely in biological systems.

Once the network is activated, it can continue operating as a dynamical system. As it is operating, the activations of its neurons keep changing until the network approaches a stable state when the continuing operation no longer changes the activations. Hopfield showed that this stable state corresponds to a local minimum of an energy function E, defined as follows:

$$E = \frac{1}{2}\sum_{i,j} w_{ij}S_iS_j + \sum_i a_iS_i$$

$$= \frac{1}{2}\mathbf{S}^T\mathbf{W}\,\mathbf{S} + \mathbf{A}^T\mathbf{S} \tag{2.36}$$

This is called an energy function, as the expression resembles the energy in mechanical systems.

The network converges to a local minimum that is closest to one of the stored vectors. This is how the stored vectors are retrieved. Connection weights are determined during the training of the Hopfield net to store the vectors. There are several methods of training the Hopfield nets to store vectors. We will only briefly describe two methods.

Let us assume that we intend to store n vectors $\mathbf{u}^k = \{u_i^k\}$; $k = 1,\ldots,n$ The method of training that Hopfield developed is shown in the following equation:

$$w_{ij} = \sum_{k=1}^{n}\left(2u_i^k - 1\right)\left(2u_j^k - 1\right) \tag{2.37}$$

The Hebbian learning rule for storing the n vectors in the Hopfield net is as follows:

$$w_{ij} = \frac{1}{n} \sum_{k=1}^{n} u_i^k u_j^k \qquad (2.38)$$

We can also write the Hebbian learning rule in matrix form where $\mathbf{W}$ is a square symmetric matrix with 0 diagonal terms and the terms of the matrix are the connection weights.

$$\mathbf{W} = \frac{1}{n} \sum_{k=1}^{n} \mathbf{U}^k \, \mathbf{U}^{k^T} \qquad (2.39)$$

$$\mathbf{W} = \left[w_{ij} \right]$$

$$\mathbf{U}^k = \left\{ u_j^k \right\}$$

The training of Hopfield net stores the n pattern vectors in its connection weights. They can be retrieved by activating the neurons with a partial or somewhat distorted version of a stored pattern vector. The activation will initiate the dynamic response of the net that will converge to the stored pattern vector most similar to the one that was used to activate it.

2.11 DISCUSSION

We have discussed mainly MLF neural networks in this chapter. This is the type of neural networks that are used in the engineering applications that will be presented in the following chapters. Supervised training and backpropagation methods are used in training MLF neural networks. There are special forms MLF NNs that are used in constitutive modeling of material behavior in computational mechanics. In the next chapter, we will present nested adaptive neural networks for modeling of material behavior. These and other neural networks that are intended to model the material behavior under loading and unloading are recurrent NNs, in a sense that the output of the neural network is used as the input in the next step.

Another special form of MLF NN that is used in some applications presented in Chapter 4 is the replicator NN. The input and output in the training data for these NNs are the same - NN learns to replicate the input in its output. The data used in training replicator NNs usually contain some underlying information, plus certain amount of variability, noise and scatter. The replicator NNs that have some form of stochastic activation functions are used as a form of data compression.

Replicator NNs have an odd number of hidden layers, and the middle hidden layer has a smaller number of nodes than the input and other hidden layers. A typical replicator NN is shown in Figure 2.24. The middle hidden layer is the data-compression layer. The lower part of the NN from the input layer to the compression layer is the encoder that compresses the data. The upper part of the NN from the compression layer to the output layer is the decoder that decompresses the data.

The activations of the compression layer contain the compressed data that mainly contain the underlying information in the training data after removing the scatter and noise.

We also need to point out the all the neural networks used in applications presented in the book are also classified as *shallow neural networks*. This distinguishes them from

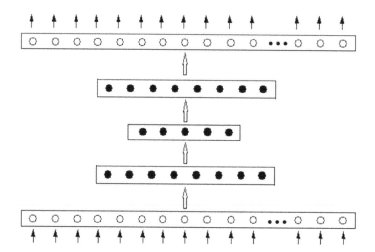

Figure 2.24 A typical replicator Neural network. The input and output in the training data are the same. The activations of the middle hidden layer are the compressed data.

deep neural networks. Deep neural networks are typically larger neural network with more hidden layers and in some cases different topology than the shallow NNs. The training of deep neural networks is also basically different than the backpropagation supervised learning that is used in shallow NNs. Deep learning methods use supervised leaning (SL), unsupervised learning (UL), and reinforcement learning (RL). Training of the deep NNs is referred to as *deep learning.* Deep NNs with deep learning have been successfully applied in image recognition, computer vision, speech recognition, and language processing. We will not cover deep learning neural networks in this book. The reader is referred to many publications in this field, including an overview in Schmidhuber (2015).

Chapter 3

Neural networks in computational mechanics

3.1 INTRODUCTION

A typical computational mechanics problem, such as finite-element analysis of solids and structural systems, includes the following tasks: discretizing the geometry of the system with a finite-element mesh; specifying the constitutive behavior of the materials in the system; applying the boundary conditions; applying the stimuli, such as forces; and, performing the analysis. In most cases, specifying the constitutive behavior of the materials is the most difficult part of the task. This is where the neural networks can play an important role. Application of neural networks in computational mechanics started in the late 1980s and the early 1990s in material modeling (Ghaboussi et al., 1990, 1991; Ghaboussi and Wu, 1998). In this chapter, we will discuss the application of neural networks in modeling of constitutive behavior of materials (Ghaboussi, 2001). First, some fundamentals of constitutive modeling are described.

Constitutive modeling is mainly based on the observed behavior in material tests. Constitutive modeling from material tests is inherently an inverse problem. If we consider the specimen in the material test as the system, then the applied stresses and measured strains can be considered as input and output. The input and output of the system are known (measured), and the constitutive model of the system needs to be determined. This is clearly an inverse problem, as discussed in Chapter 1, Section 1.5, and will be discussed further in Chapter 4. In material modeling, we normally do not solve this inverse problem directly. Rather than directly determining the model for the material behavior, we normally use a combination of principles (i.e., conservation laws, objectivity and frame indifference) and assumptions or idealizations (e.g., isotropy, elasticity, plasticity, yield surface, normality rule, and hardening) and so on. Using these principles and assumptions leads to some mathematical model, and the objective of material testing reduces to determining the parameters of the mathematical material model. Almost all material tests in practice are performed to determine the parameters of a specific mathematical material model that we have already chosen, and which satisfies the fundamental principles and our assumptions.

When we use neural networks, we train a neural network to learn the material behavior. In doing so, we are directly solving the inverse problem of determining the constitutive material model. The trained neural network can then be used in finite-element simulation, similar to the way mathematical material models are used. There are many important issues in using the neural networks in material modeling that will be discussed in this chapter and Chapter 5.

3.2 NEURAL NETWORKS IN MODELING CONSTITUTIVE BEHAVIOR OF MATERIAL

Modeling of the nonlinear constitutive behavior of materials is an important part of the numerical simulation of structural systems with finite-element or finite-difference methods. Neural networks can be used to model the nonlinear constitutive behavior of materials. Mathematically based constitutive models, such as plasticity, arrive at an incremental relation between the increments of strains and increments of stresses, as in the following equation:

$$\Delta\sigma = \mathbf{C}^{ep}\Delta\varepsilon \tag{3.1}$$

In this equation, $\mathbf{C}^{ep}$ is the elastoplastic constitutive matrix.

Neural networks can also be trained to represent the nonlinear constitutive behavior of materials. In its simplest form, a neural network material model that can be used in a finite-element analysis takes strains (or strain increments) as input and gives the stresses (or stress increments) as output. In the earliest applications, neural networks were used in this form. In this case, we simply replace the mathematical model with a neural network. However, neural networks can be used in far more effective ways, as will be discussed later.

In the first applications in computational mechanics (Ghaboussi et al., 1990, 1991; Wu, 1991) neural networks were used to model the behavior of plain concrete in a two-dimensional plane stress condition under monotonic loading. Existing experimental results were used to train a neural network. Because of the nonlinearity and path dependence, the current stresses and strains and the strain increments were used as input to the neural network with the stress increments as the output, as shown in Figure 3.1.

$$\Delta\sigma = \text{NN}\,[\sigma,\ \varepsilon,\ \Delta\varepsilon : 6\,|\,30\,|\,30\,|\,2] \tag{3.2}$$

This type of neural network material model was described as a *one-point scheme*; it uses the current state of stresses and strains as part of the input. This neural network was

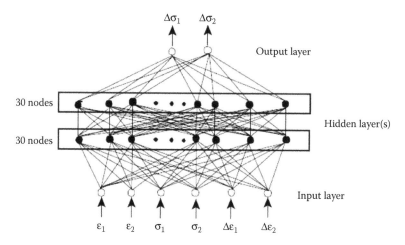

Figure 3.1 First application of neural network in modeling the behavior of plain concrete in two-dimensional stress–strain states.

able to learn the material behavior in monotonic loading experiments. Typically, a neural network is trained on some experimental results and then tested on completely different experimental results to verify that it has learned the underlying material behavior with sufficient accuracy and that it can generalize to other stress paths that were not present in the training data.

The input to this particular neural network is not sufficient to describe the material behavior in loading and unloading. Materials like metals, concrete, rocks, and geomaterials undergo microstructural changes and their loading, unloading, and reloading behavior can be significantly different. For the neural network to learn the loading and unloading behavior of the material, three history points were provided as input. This *three-point scheme*, described by the following equation, was first used for modeling plain concrete material behavior under uniaxial cyclic loading (Ghaboussi et al., 1991; Wu, 1991).

$$\Delta\sigma_{j+1} = NN\ [\sigma_{j-2},\ \varepsilon_{j-2},\ \sigma_{j-1},\ \varepsilon_{j-1},\ \sigma_j,\ \varepsilon_j,\ \Delta\varepsilon_{j+1}:\quad] \tag{3.3}$$

Since this equation is generic and it does not represent a specific neural network architecture, the number of nodes in layers is not specified in this equation.

Before continuing the discussion of modeling of the path dependence of the material constitutive properties with neural networks, in the following few sections, we will discuss the nested adaptive neural networks (NANN) that play an important role in constitutive modeling with neural networks.

3.3 NESTED STRUCTURE IN ENGINEERING DATA

3.3.1 Introduction

Certain inherent structure or structures may exist within the engineering data, such as the data being used to train a neural network material model. These data structures may play an important role in training of neural networks. It is important to point out that these data structures are usually of no consequence in mathematically based methods of analysis. As a result, the users of the data are seldom aware of any structures in engineering data. In fact, similar structures also exist in many nonengineering data.

Taking advantage of the structures in the training data often results in efficient neural network architecture. In this section, we will first describe the structures that may exist within the data and how these structures can be reflected in the internal structure of the neural networks. One important type is the nested structure. We will describe the NANNs that take advantage of the nested structure in the data. We will present some examples of the NANNs in modeling of the constitutive behavior of materials.

3.3.2 Nested structure in training data

Nested data structures can occur in many different types of data, and it may occur at many different levels. The simplest form of nested structure exists is the spatial dimensions. Assume that a certain observation of a natural or man-made phenomenon is made and the terms of vector **Y**, quantities y_1 through y_N, are measured. Also assume that these measurements can be made in one-, two-, and three-dimensional cases. Also assume that the same measurements can be made as a function of time or a timelike variable. For example, the measurements can be of temperature, pressure and maybe some other quantities, measured at different spatial locations and also as a function of time. The measurements are almost always at discrete space and time points or they are discretized.

There are potentially four levels of nested structure in this type of data. A general three-dimensional time-dependent problem can be made into a time-independent problem by restricting it to a specific value of time. Both the time-dependent and time-independent three-dimensional problems can further be restricted to two-dimensional and one-dimensional problems. The nested structure arises from the dimensionality and time dependence of the measured data. The one-dimensional version of problems is the subset of the two-dimensional version of the same problem. Similarly, the two-dimensional problem itself is a subset of the three-dimensional problem. The time-independent problems in one-, two-, and three dimensions are subsets of the time-dependent problems.

The dependence of the vector $\mathbf{Y}$ on the spatial coordinates can be expressed mathematically as functions of those spatial coordinates.

$$\begin{cases} \mathbf{Y}_1 = \mathbf{f}_1\left(x_1\right) \\ \mathbf{Y}_2 = \mathbf{f}_2\left(x_1, x_2\right) \\ \mathbf{Y}_3 = \mathbf{f}_3\left(x_1, x_2, x_3\right) \end{cases} \tag{3.4}$$

We will simply refer to $\mathbf{f}_1$, $\mathbf{f}_2$, and $\mathbf{f}_3$, as the one-, two-, and three-dimensional functions. They can be thought of as being members of the function spaces $\mathbf{F}_1$, $\mathbf{F}_2$, and $\mathbf{F}_3$.

$$\begin{cases} \mathbf{f}_1 \in \mathbf{F}_1 \\ \mathbf{f}_2 \in \mathbf{F}_2 \\ \mathbf{f}_3 \in \mathbf{F}_3 \end{cases} \tag{3.5}$$

The nested structure of the data can be expressed in terms of the function spaces. The function space $\mathbf{F}_1$ is a subspace of the function space $\mathbf{F}_2$, which in turn is a subspace of the function space $\mathbf{F}_3$.

$$\mathbf{F}_1 \subset \mathbf{F}_2 \subset \mathbf{F}_3 \tag{3.6}$$

These relationships can be further clarified, if we consider that the two-dimensional function $\mathbf{f}_2$ can be determined from the three-dimensional function $\mathbf{f}_3$ by fixing the value of one of the coordinates. Similarly, $\mathbf{f}_1$ can be obtained from $\mathbf{f}_2$ by fixing the value of one of the coordinates.

$$\begin{cases} \mathbf{f}_1\left(x_1\right) = \mathbf{f}_2\left(x_1, a\right) \\ \mathbf{f}_2\left(x_1, x_2\right) = \mathbf{f}_3\left(x_1, x_2, a\right) \end{cases} \tag{3.7}$$

In these equations, "a" is a fixed valued parameter.

These observations can be extended to the time-dependent multidimensional problems as well. If the vector $\mathbf{Y}$ is measured as a function of time as well, then we will have the following equations:

$$\begin{cases} Y_1 = \bar{f}_1(x_1, t) \\ Y_2 = \bar{f}_2(x_1, x_2, t) \\ Y_3 = \bar{f}_3(x_1, x_2, x_3, t) \end{cases} \tag{3.8}$$

These functions can be thought of as being members of the following function spaces:

$$\begin{cases} \bar{f}_1 \in \bar{F}_1 \\ \bar{f}_2 \in \bar{F}_2 \\ \bar{f}_3 \in \bar{F}_3 \end{cases} \tag{3.9}$$

The following nested relationships also exist between the functions spaces:

$$\begin{cases} F_1 \subset \bar{F}_1 \\ F_2 \subset \bar{F}_2 \\ F_3 \subset \bar{F}_3 \end{cases} \tag{3.10}$$

The nested structure can be generalized to an n-dimensional case. The measured terms of the vector Y may be a function of n variables. Similar functional relationships may exist for different values of n.

$$Y_n = f_n(x_1,\ldots, x_n); \quad n = 1, 2,\ldots \tag{3.11}$$

The function spaces, to which these functions belong, have the following nested structure:

$$F_n \subset F_{n+1} \tag{3.12}$$

This nested structure can be exploited in developing and training the neural networks to represent these functions.

3.3.3 Nested structure in constitutive behavior of materials

Modeling of the constitutive behavior of materials deals with relations between stresses and strains for materials. Stresses and strains are tensor quantities, and they have six independent components in the general three-dimensional states. In computational mechanics, the stresses and strains are used in vector form. These vectors contain the independent components of the stress and strain tensor.

Data on the constitutive behavior of materials come from the material tests in the laboratory. In this section, we will discuss the nested structure of the data from material tests, which can be used in training of the neural networks to represent the material behavior.

In the case of material data, there are several types of nested structures. At this point, we will consider the type of the nested structure related to the dimensionality. In a later section, we will discuss other types of nested structure in material data, such as those resulting from the history or path dependence in material behavior.

A rate form of the constitutive model relates the rate of stresses to the rate of strains and the current state of stresses and strains, as expressed in the following equation:

$$\dot{\sigma}_j = f_j \left(\sigma_j,\ \varepsilon_j,\ \dot{\varepsilon}_j \right) \quad j = 1, 3, 4, 6$$

$$\begin{cases} \sigma = \text{stress vector} \\ \varepsilon = \text{strain vector} \end{cases} \tag{3.13}$$

and j is the number of terms in the stress and strain vectors. Material data can come from a number of different material tests, and they vary in the number of independent stress and strain terms.

In a uniaxial material test, there is only one component in each of the stress and the strain vectors.

$$\begin{cases} \sigma = \left\{ \sigma_{11} \right\} \\ \varepsilon = \left\{ \varepsilon_{11} \right\} \end{cases} \tag{3.14}$$

Uniaxial tests can be performed in two different ways, as shown in Figure 3.2. In uniaxial stress test, a bar is subjected to tension or compression. In a uniaxial strain test, the material is laterally confined and is usually subjected to compression. Uniaxial strain tests are rare, and they are mainly performed on geomaterials, such as soils and rocks. Both types of test can be performed either as stress controlled, where the stress is applied, and the strain is measured, or as strain controlled, where the strain is applied, and the stress is measured.

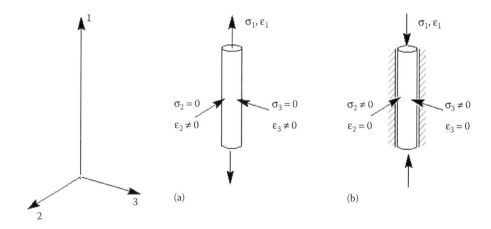

Figure 3.2 Uniaxial material tests: (a) uniaxial stress test, (b) uniaxial strain test laterally constrained.

In a one-dimensional stress test, σ_{11} is the uniaxial stress and ε_{11} is the uniaxial strain. All the other stresses and shear strains are zero; $\sigma_{22} = \sigma_{33} = \sigma_{12} = \sigma_{23} = \sigma_{13} = \varepsilon_{12} = \varepsilon_{23} = \varepsilon_{13} = 0$. The two lateral normal strain components, ε_{22} and ε_{33}, are nonzero but normally have no effect on the uniaxial stress–strain relations.

In a uniaxial strain test, again σ_{11} and ε_{11} are the uniaxial stress and strain. All the other strains and shear stresses are zero; $\varepsilon_{22} = \varepsilon_{33} = \varepsilon_{12} = \varepsilon_{23} = \varepsilon_{13} = \sigma_{12} = \sigma_{23} = \sigma_{13} = 0$. The two lateral normal stress components, σ_{22} and σ_{33}, are nonzero.

In two-dimensional states, there are three independent components of stresses and strains:

$$\begin{cases} \sigma = \{\sigma_{11}, \ \sigma_{22}, \ \sigma_{12}\} \\ \varepsilon = \{\varepsilon_{11}, \ \varepsilon_{22}, \ \varepsilon_{12}\} \end{cases} \tag{3.15}$$

Two-dimensional problems are either plane stress or plane strain, as shown in Figure 3.3. The out-of-plane shear stresses and strains are zero; $\sigma_{23} = \sigma_{13} = \varepsilon_{23} = \varepsilon_{13} = 0$. In the case of two-dimensional plane stress, the out-of-plane normal stress is zero, $\sigma_{33} = 0$, whereas the corresponding strain, ε_{33}, is nonzero. The opposite is true in the case of the two-dimensional plane-strain, where the out-of-plane strain is zero, $\varepsilon_{33} = 0$, and the corresponding stress component, σ_{33}, is nonzero.

In axisymmetric states, shown in Figure 3.4, there are four nonzero components of stress and strain.

$$\begin{cases} \sigma = \{\sigma_{rr}, \ \sigma_{\theta\theta}, \ \sigma_{zz}, \ \sigma_{rz}\} \\ \varepsilon = \{\varepsilon_{rr}, \ \varepsilon_{\theta\theta}, \ \varepsilon_{zz}, \ \varepsilon_{rz}\} \end{cases} \tag{3.16}$$

Polar coordinates are used on the plane normal to the axis of the sample. The coordinate system $(1, 2, 3)$ corresponds to (r, θ, z).

In the most general case of three-dimensional problems, all the six components of stresses and strain are present, as shown in Figure 3.5.

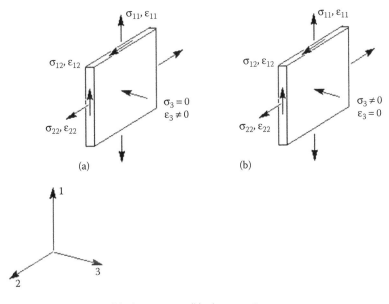

Figure 3.3 Two-dimensional states: (a) plane stress, (b) plane strain.

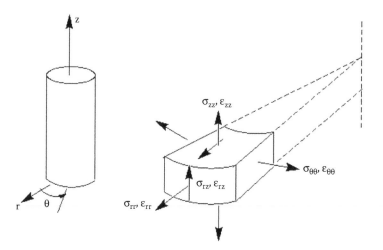

Figure 3.4 State of stresses and strains in axisymmetric problems.

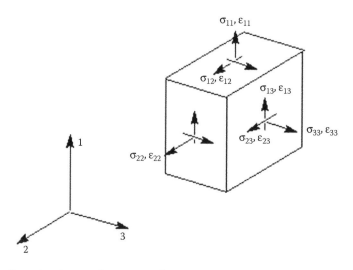

Figure 3.5 Three-dimensional state of stresses and strains.

There are three normal stress and strain components and three shear stress and strain components.

$$\begin{cases} \sigma = \{\sigma_{11}, \sigma_{22}, \sigma_{33}, \sigma_{12}, \sigma_{23}, \sigma_{31}\} \\ \varepsilon = \{\varepsilon_{11}, \varepsilon_{22}, \varepsilon_{33}, \varepsilon_{12}, \varepsilon_{23}, \varepsilon_{31}\} \end{cases} \tag{3.17}$$

The constitutive functions $\mathbf{f}_j$ can be considered to be members of function spaces $\mathbf{F}_j$.

$$\mathbf{f}_j \in \mathbf{F}_j \tag{3.18}$$

These function spaces clearly have the nested structure, as shown in the following equation:

$$F_1 \subset F_3 \subset F_4 \subset F_6 \tag{3.19}$$

This equation states that one-dimensional constitutive behavior is a subset of the constitutive behavior in two-dimensional problems, which in turn is a subset of the constitutive behavior in axisymmetric problems, which in turn is a subset of the constitutive behavior in three-dimensional state of stresses and strains.

There is also nested structure in standard triaxial tests and true triaxial tests, where shear stresses and shear strains are zero. The state of stresses and strains in these tests, which are usually performed on geomaterials under compression, are shown in Figure 3.6.

In the standard triaxial tests, which are performed on geomaterials, an axisymmetric sample is subjected to radial all around stress, σ_r, and a uniaxial stress, σ_a. The material behavior in a standard triaxial test is characterized by two independent components of stresses and strains, represented by the stress vector $\boldsymbol{\sigma} = \{\sigma_a, \sigma_r\} = \{\sigma_{11}, \sigma_{22} = \sigma_{33}\}$ and the corresponding strain vector $\boldsymbol{\varepsilon} = \{\varepsilon_a, \varepsilon_r\} = \{\varepsilon_{11}, \varepsilon_{22} = \varepsilon_{33}\}$. The material behavior in true triaxial test is characterized by three independent components of stresses and strains, represented by the stress vector $\boldsymbol{\sigma} = \{\sigma_1, \sigma_2, \sigma_3\} = \{\sigma_{11}, \sigma_{22}, \sigma_{33}\}$ and the corresponding strain vector $\boldsymbol{\varepsilon} = \{\varepsilon_1, \varepsilon_2, \varepsilon_3\} = \{\varepsilon_{11}, \varepsilon_{22}, \varepsilon_{33}\}$. All the shear stresses and strains in both types of tests are zero.

The constitutive function $\bar{f}$ has been used to denote the standard triaxial and true triaxial states. These functions can also be thought of as being members of function spaces, as shown below:

$$\dot{\sigma}_j = \bar{f}_j\left(\sigma_j, \varepsilon_j, \dot{\varepsilon}_j\right) \quad j = 2, 3 \tag{3.20}$$

$$\bar{f}_j \in \bar{F}_j \tag{3.21}$$

These function spaces also have a nested structure, as shown in the following equations:

$$F_1 \subset \bar{F}_2 \subset \bar{F}_3 \subset F_6 \tag{3.22}$$

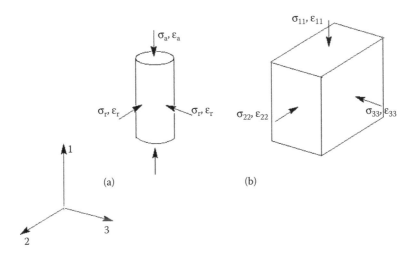

Figure 3.6 Compressive tests on geomaterials: (a) standard triaxial test state, (b) true triaxial test state.

The nested structure of the material data should be reflected in the neural networks that are trained with those data. The neural networks trained with the results of higher dimensional tests should also represent the lower dimensional stress states. Other types of nested structure in the material data will be discussed later. Some examples from the work of the author will also be presented in Section 3.6.

3.4 NESTED ADAPTIVE NEURAL NETWORKS

The nested structure of the data can be exploited in training neural networks, such that the trained neural networks also have a nested structure. The basic principle behind the nested neural networks can be stated as follows: A neural network trained with the data at a certain level should also represent the relationships present in the lower level datasets.

Consider two neural networks that are trained with dataset $[\mathbf{Y}_n: x_1,..., x_n]$ and $[\mathbf{Y}_{n+1}: x_1,..., x_{n+1}]$, shown in the following equations:

$$\begin{cases} \mathbf{Y}_n = NN_n\big[x_1,..., \ x_n; \quad \big] \\ \mathbf{Y}_{n+1} = NN_{n+1}\big[x_1,..., \ x_{n+1}; \quad \big] \end{cases} \tag{3.23}$$

The same nested structure that exists in the data and the underlying relations present in the data should also exist in the resulting trained neural networks. By this, we mean that the lower level neural network should be a subset of the higher level neural network, as expressed by the following equation:

$$NN_n \subset NN_{n+1} \tag{3.24}$$

Perhaps the use of the symbol for subset in the context of neural networks is somewhat unorthodox. The equation symbolically represents the fact that the information and relationships contained in the connection weights of the lower level neural network are a subset of those of the higher level neural network.

In this chapter, we confine ourselves to multilayered feedforward (MLF) neural networks. Equation 3.24 implies that we should be able construct n+1 from the lower level neural network n by adding an input node to represent x_{n+1}, and adding nodes in the hidden layers as needed, as shown in Figure 3.7. The requirement for satisfying Equation 3.24 and what it represents is that the connections from the added nodes to the nodes in NN_n be one-way connections. It can be visually verified that this type of one-way connections will satisfy the following equation:

$$NN_n\big[x_1,..., \ x_n \quad ; \quad \big] = NN_{n+1}\big[x_1,..., \ x_n, 0 \quad ; \quad \big] \tag{3.25}$$

Providing zero input for x_{n+1} will eliminate the effect of the additional connection weights, and the output will be the same as that of the lower level neural network.

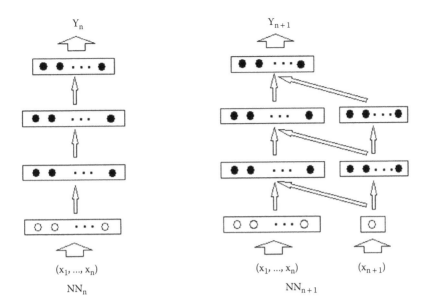

Y_n

Y_{n+1}

$(x_1, ..., x_n)$

NN_n

$(x_1, ..., x_n)$

(x_{n+1})

NN_{n+1}

Figure 3.7 Schematics of constructing a higher level nested neural network from a lower level neural network.

A nonzero value of the input for x_{n+1} will have a fixed effect on the output of the neural network. This effect will reflect the training history of the higher level neural network with the nonzero values of x_{n+1}.

The training of nested neural networks also requires some attention. It is assumed that NN_n is already a trained neural network. When the additional node is added to create NN_{n+1}, some new connections are also created. During the additional training required for determining the new connection weights, the existing connections of NN_n must remain frozen. This requirement is needed to satisfy Equation 3.25, to have the ability to retrieve the lower level neural network when the input value for the node x_{n+1} is set to zero.

The observations on the nested data structure and their relevance to the neural network architecture form the basis of the NANNs. As shown in Figure 3.8, every move to one higher level, involves adding a new module. The NANNs have a modular structure, starting from a neural network at the lowest level that is called the *base module*. The additional modules are added with one-way connections, as shown in Figure 3.8. This process is open-ended, and as many modules can be added as needed to represent a problem adequately.

We make two additional observations. The case shown in Figure 3.7 is a special case. In general, the added modules may have more than one input node. The input to the added modules can also be vector quantities. Also, the added modules may have their own output nodes. The added output nodes for the added modules in Figure 3.8 have been placed in boxes to indicate that in some cases they may not exist. The general case shown schematically in Figure 3.8 has both of these features.

At each stage when a new module is added, and new connections are created, additional training is required for those new connections weights. The connection weights of the lower level nested neural network are frozen, whereas the training takes place to determine the weights for the new connections.

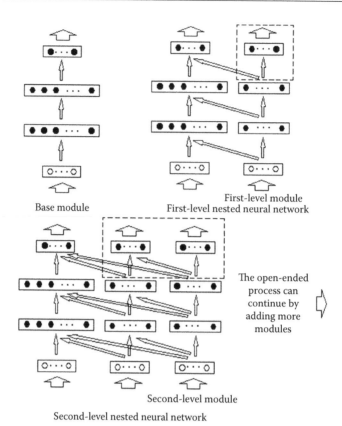

Figure 3.8 Process of progressively creating higher level nested adaptive neural networks.

3.5 PATH DEPENDENCE AND HYSTERESIS IN CONSTITUTIVE BEHAVIOR OF MATERIALS

Nonlinear constitutive behavior of materials is inherently path dependent. The relationship between the strain rate and the stress rate are not only dependent on the current state of stresses and strains, they also depend on the immediate past history of stresses and strains. Usually, in material modeling, the immediate past history is available at discrete points. In material experiments, the discrete points correspond to the applied increments of stresses or increments of strains. In computational mechanics, stresses and strains are computed in incremental analyses. A typical constitutive model is shown in the following equation, where n is the increment number and k is the number of history point.

$$\dot{\sigma}_n = f_{nk}\left(\dot{\varepsilon}_n, \sigma_n, \varepsilon_n, \sigma_{n-1}, \varepsilon_{n-1}, \ldots, \sigma_{n-k}, \varepsilon_{n-k}\right)$$

$$\begin{cases} n = 1, 2, \ldots \\ k = 0, \ 1, 2, \ldots \end{cases} \tag{3.26}$$

In practice, either in experiments or in computational mechanics, often the incremental form of the constitutive model is used.

$$\Delta\sigma_n = f_{nk}\left(\Delta\varepsilon_n, \sigma_n, \varepsilon_n, \sigma_{n-1}, \varepsilon_{n-1}, \ldots, \sigma_{n-k}, \varepsilon_{n-k}\right)$$

$$\begin{cases} n = 1, 2, \ldots \\ k = 0, 1, 2, \ldots \end{cases} \tag{3.27}$$

Neural networks can be used to learn the incremental form of the constitutive model, and NANN can be used to empirically determine the value of k in the process. Figure 3.9 shows the process of training a typical uniaxial constitutive model with NANN

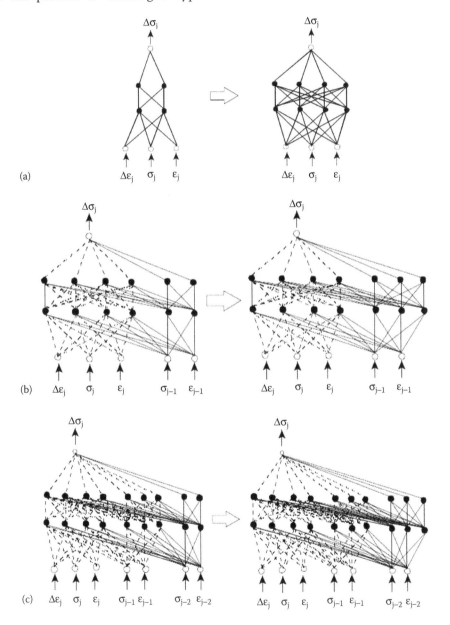

Figure 3.9 The evolution and training of a typical nested adaptive neural network material model with two history point modules: (a) the base module and its adaptive training, (b) addition and adaptive training of the first history point module, (c) addition and adaptive training of the second history point module.

with two history points. These neural networks can be shown in compact form in the following expressions.

These equations clearly show the evolution of the NANN and the network architecture, as well as the process of their adaptive training.

$$
\begin{cases}
\Delta\sigma_n = NN_0\big[\Delta\varepsilon_n, \sigma_n, \varepsilon_n \; ; \; 3\,|\,2-4\,|\,2-4\,|\,1\big] \\[2mm]
\Delta\sigma_n = NN_1\left[\begin{matrix} (\Delta\varepsilon_n, \sigma_n, \varepsilon_n)(\sigma_{n-1}, \varepsilon_{n-1}) \; ; \\[1mm] 3,2\,|\,2-4,\,2\,-\,3\,|\,2-4,\,2\,-\,3\,|\,1 \end{matrix}\right] \\[4mm]
\Delta\sigma_n = NN_2\left[\begin{matrix} (\Delta\varepsilon_n, \sigma_n, \varepsilon_n)(\sigma_{n-1}, \varepsilon_{n-1})(\sigma_{n-2}, \varepsilon_{n-2}) \; ; \\[1mm] 3,2,2\,|\,2-4,\,2-3,\,2-3\,|\,2-4,\,2-3,\,2-3\,|\,1 \end{matrix}\right]
\end{cases}
\tag{3.28}
$$

The base module starts with two nodes in each hidden layer. During the adaptive training, the number of nodes in each hidden layer increases to four. Similarly, when a new module is added, initially two nodes are assigned to each hidden layer. During the adaptive training, the number of nodes in the hidden layers of the added module increase and as many additional nodes as needed are added.

Path dependence of the material behavior is clearly demonstrated in the hysteretic behavior when the material is subjected to cyclic stresses or strains. Hysteretic behavior of materials can be modeled by using history points, as will be demonstrated in the next section with an example of plain concrete subjected to cyclic compressive stresses.

Another form of modeling of the hysteretic behavior of materials with neural networks was reported in Yun et al. (2008a). This method uses the strain energy and the change in the strain energy as part of the input to the neural network. A typical neural network material model with this method is shown in the following equation:

$$
\sigma_n = NN\big[\varepsilon_n, \sigma_{n-1}, \varepsilon_{n-1}, \xi_{n-1}, \Delta\eta_\varepsilon \; ; \quad \big]
$$

$$
\begin{cases}
\xi_{n-1} = \sigma_{n-1}\,\varepsilon_{n-1} \\[2mm]
\Delta\eta_\varepsilon = \sigma_{n-1}\big(\varepsilon_n - \varepsilon_{n-1}\big)
\end{cases}
\tag{3.29}
$$

In the equation, ξ is the strain energy and $\Delta\eta_\varepsilon$ is the change in the strain energy in strain-controlled problems. An alternative form of the change in the strain energy can also be used in mainly stress controlled problems. These two forms of the strain energy change are equally effective in most problems. Further details of this methodology are available in the publication cited earlier.

This method was mainly applied in modeling of beam–column connections subjected dynamic cyclic loads. Some examples of the application of this method will be presented in Chapter 5.

3.6 CASE STUDIES OF APPLICATION OF NESTED ADAPTIVE NEURAL NETWORKS IN MATERIAL MODELING

3.6.1 Uniaxial cyclic behavior of plain concrete

As an example of a uniaxial constitutive model, we will present the case of monotonic and cyclic behavior of plain concrete in compression. The example is taken from the doctoral dissertation of Dr. Mingfu Zhang (Zhang, 1996).

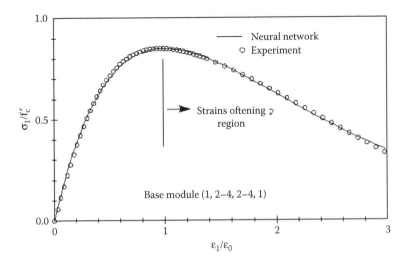

Figure 3.10 Modeling of the monotonic uniaxial behavior of plain concrete under compressive loading. (From Zhang, M., Determination of neural network material models from structural tests, PhD thesis, Department of Civil and Environmental Engineering, University of Illinois at Urbana-Champaign, Urbana, IL, 1996.)

Figure 3.10 shows the experimental results (Smith and Young, 1955) and the neural network response for the uniaxial monotonic behavior of concrete in compression. The compressive stress in plain concrete reaches a maximum and the declining portion beyond that is referred to as the strain softening region.

The neural network used in Figure 3.10 is shown in the following equation for increment number n, where $\sigma = \sigma_1/f'$ and $\varepsilon = \varepsilon_1/\varepsilon_0$ are the normalized stresses and strains, f' is the uniaxial compressive strength of concrete and ε_0 is the uniaxial strain at peak stress.

$$\sigma_n = NN_0 \left[\varepsilon_n \; ; 1 \; | 2-4 | 2-4 | \; 1 \right] \tag{3.30}$$

This neural network served as the base module for the higher level NANNs needed for modeling of the cyclic behavior of concrete.

Under uniaxial cyclic stresses, plain concrete degrades. As a result, the peak stress at each cycle is lower than at the previous cycle, as shown in Figure 3.11. This type of behavior is very complex and obviously dependent on the past history. At any point on the curve, just the values of stress and strain are not sufficient to describe the material behavior. The stresses and strains at several points in the immediate past history are needed to describe the material behavior. The question is how many points along the immediate past history are sufficient to describe the material behavior with a reasonable degree of accuracy. Again, the NANNs can be used to empirically determine the number of history points needed.

The experiments were performed by applying incremental stresses. As a result, the experimental curves shown in Figure 3.11 are a collection of discrete points. These discrete points were used to create the training dataset for the neural network models of the material behavior. The objective is for the neural network to learn the relationship shown in the following

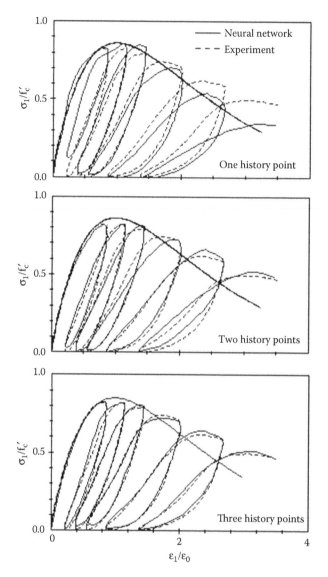

Figure 3.11 Training results of concrete under uniaxial monotonic and cyclic compression loading with three NANNs shown in Equation 3.32: NN1, with one history point; NN2 with two history points; and NN3, with three history points. (From Zhang, M., *Determination of neural network material models from structural tests*, PhD thesis, Department of Civil and Environmental Engineering, University of Illinois at Urbana-Champaign, Urbana, IL, 1996.)

equation, where $\sigma = \sigma_1/f'$ and $\varepsilon = \varepsilon_1/\varepsilon_0$ are the normalized stresses and strains and n is the increment number.

$$\sigma_n = f\left(\varepsilon_n, \ \sigma_{n-1}, \ \varepsilon_{n-1}, \ldots, \ \sigma_{n-k}, \ \varepsilon_{n-k}\right) \tag{3.31}$$

The value of k is determined by testing the accuracy at each level of NANN.

The base module in these NANNs was the neural network trained with the monotonic loading, shown in Equation 3.30. Up to three history points were added: first level NN_1 with one history point; second level NN_2 with two history points; and finally arriving at the third level NANN NN_3 with three history points. The third level NANN, NN_3, was judged to have learned the constitutive behavior with reasonable accuracy. The series of NANNs generated by adding history points are shown in the following equation:

$$\left\{ \begin{array}{l} \sigma_n = NN_1 \begin{bmatrix} (\varepsilon_n), (\sigma_{n-1}, \varepsilon_{n-1}) \ ; \\ 1, 2 \ | 2-4, 2-12 | 2-4, 2-12 | \ 1 \end{bmatrix} \\[4ex] \sigma_n = NN_2 \begin{bmatrix} (\varepsilon_n), (\sigma_{n-1}, \varepsilon_{n-1}), (\sigma_{n-2}, \varepsilon_{n-2}) \ ; \\ 1, 2, 2 \ | 2-4, 2-12, 2-10 | \\ 2-4, 2-12, 2-10 | \ 1 \end{bmatrix} \\[6ex] \sigma_n = NN_3 \begin{bmatrix} (\varepsilon_n), (\sigma_{n-1}, \varepsilon_{n-1}), (\sigma_{n-2}, \varepsilon_{n-2}), (\sigma_{n-3}, \varepsilon_{n-3}) \ ; \\ 1, 2, 2, 2 \ | 2-4, 2-12, 2-10, 2-9 | \\ 2-4, 2-12, 2-10, 2-9 | \ 1 \end{bmatrix} \end{array} \right. \qquad (3.32)$$

When each history module was added, it initially had two nodes in each of its hidden layers. Additional nodes were added to the hidden layers during the training as needed, according to the adaptive architecture method described in Section 2.8. In Equation 3.32, we can see the adaptive training of the base module and each of the history points. The base module started with two nodes in each hidden layer and ended with four nodes. The first history point in NN_1 started with 2 nodes in each hidden layer and adaptively increased to 12 nodes. The second history point in NN_2 also started with 2 nodes in each hidden layer and adaptively increased to 10 nodes. Finally, the third history point in NN_3 started with 2 nodes in each hidden layer and adaptively increased to 9 nodes.

The results of a series of NANNs trained to learn the uniaxial cyclic behavior of plain concrete are shown in Figure 3.11. The results of the testing of the trained neural networks on a different experiment that was not part of the training data are shown in Figure 3.12.

Figures 3.11 and 3.12 clearly show that as more history points are added, the NANN is able to learn the cyclic behavior of the material with increasing accuracy.

It is important to note that the appropriate testing of these neural networks, which was used in Figures 3.11 and 3.12, requires that the input to the history nodes be the output values generated by the neural network in the previous steps, not the actual experimental values; we can consider this process as *feedback* from output to input. At each step, only the value of ε_n is taken from the experiment results. This is the procedure that has been used in the normal applications of the trained neural networks as material models in computational mechanics where the experimental data will not be available.

3.6.2 Constitutive model of sand in triaxial state

Triaxial tests on sand are performed both in drained and in undrained condition. In drained condition, the pore water pressure, u, remains zero, whereas in undrained condition pore

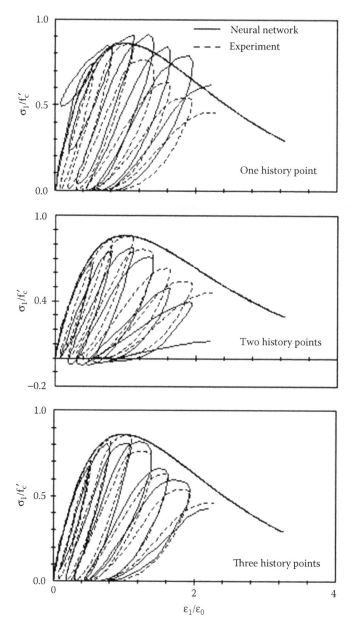

Figure 3.12 Testing results of concrete under uniaxial monotonic and cyclic compression loading with three NANNs shown in Equation 3.32: NN1, with one history point; NN2, with two history points; and NN3, with three history points. (From Zhang, M., Determination of neural network material models from structural tests, PhD thesis, Department of Civil and Environmental Engineering, University of Illinois at Urbana-Champaign, Urbana, IL, 1996.)

water pressure is nonzero and can change during the test. NANNs were used in Ghaboussi and Sidarta (1998), Sidarta (2000) to model both the drained and the undrained behavior of sand in triaxial tests. Three-level NANN, consisting of a base module NN_0 and two higher levels with history point modules NN_1 and NN_2 were used. These NANNs are given in the following expressions, and they are shown in Figure 3.13.

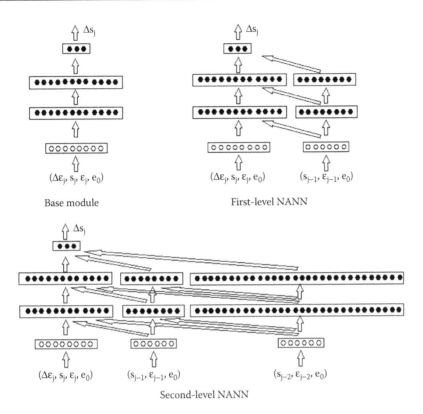

Figure 3.13 The sequence of developing nested adaptive neural networks for modeling of drained and undrained behavior of sand. (From Ghaboussi, J. and Sidarta, D.E., *Int. J. Computer Geotechnics*, 22, 29–51, 1998.)

$$
\begin{cases}
\Delta s_j = NN_0\left[(\Delta\varepsilon_j,\ s_j,\ \varepsilon_j,\ e_0);\ 8\ |\ 2-12\ |\ 2-12\ |\ 3\right] \\[2ex]
\Delta s_j = NN_1\begin{bmatrix} (\Delta\varepsilon_j,\ s_j,\ \varepsilon_j,\ e_0),\ (s_{j-1},\ \varepsilon_{j-1},\ e_0); \\[1ex] 8,\ 6\ |\ 2-12,\ 2-8\ |\ 2-12,\ 2-8\ |\ 3 \end{bmatrix} \\[3ex]
\Delta s_j = NN_2\begin{bmatrix} (\Delta\varepsilon_j,\ s_j,\ \varepsilon_j,\ e_0),\ (s_{j-1},\ \varepsilon_{j-1},\ e_0),\ (s_{j-2},\ \varepsilon_{j-2},\ e_0); \\[1ex] 8,\ 6,\ 6\ |\ 2-12,\ 2-8,\ 2-30\ |\ 2-12,\ 2-8,\ 2-30\ |\ 3 \end{bmatrix}
\end{cases}
\tag{3.33}
$$

In these equations $s = \{\sigma,\ u\}$, $\sigma = \{p',\ q\}$, u = pore pressure, $\varepsilon = \{\varepsilon_v,\ \varepsilon_d\}$, e_0 = void ratio, $p' = (\sigma'_{11} + 2\sigma'_{33})/3$ = mean effective stress, $q = \sigma_{11} - \sigma_{33}$ = deviatoric stress, $\varepsilon_v = \varepsilon_{11} + 2\varepsilon_{33}$ = volumetric strain, $\varepsilon_d = \varepsilon_{11} - \varepsilon_{33}$ = deviatoric strain, σ_{11} = axial stress, $\sigma_{33} = \sigma_{22}$ = lateral stresses, ε_{11} = axial strain, $\varepsilon_{33} = \varepsilon_{22}$ = lateral strains.

The details of training of the neural networks and the experimental results used in the training can be found in Ghaboussi and Sidarta (1998), Sidarta (2000). All the modules were trained adaptively. Initially they started with two nodes in each hidden layer and during the training additional nodes were added to the hidden layers as needed.

In the previous section, we briefly discussed the appropriate method of evaluating the performance of NANNs. We have the same issue with the NANNs in section, and in fact, with all the NANNs. For evaluating the performance of the trained neural network material models, they can be used in two different ways, differing only in the way the current state of stresses, strains and pore pressures and their history points are specified as input to the neural network. One procedure is to use the experimental data for the current state and history states of stresses, strains and pore pressures at all stages of the loading. The trained neural network material models usually perform very well when used according to this procedure. However, they cannot be used in this manner in a computational environment when experimental data are not available. A more realistic method of evaluating the performance of the trained neural networks is to use them in the same way that they will be used as a material model in a computational environment, such as a finite-element analysis. In these circumstances, the output of the neural network must be used to accumulate the stresses and pore pressures to be used as input to the neural network in the future steps. The trained NANNs have been evaluated by using the second procedure, where the output of the neural network is used as the input at future time steps.

Some selected results are shown in Figures 3.14 and 3.15. Comparison of the neural network results and experimental results for the drained tests is shown in Figure 3.14 for initial void ratio of 0.78 and a range of initial effective pressures from 1.0 to 40.10 kg/cm².

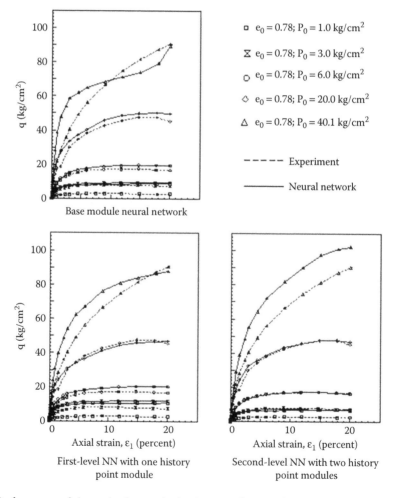

Figure 3.14 Performance of the trained nested adaptive neural network on drained triaxial compression tests. (From Ghaboussi, J. and Sidarta, D.E., *Int. J. Computer Geotechnics*, 22, 29–51, 1998.)

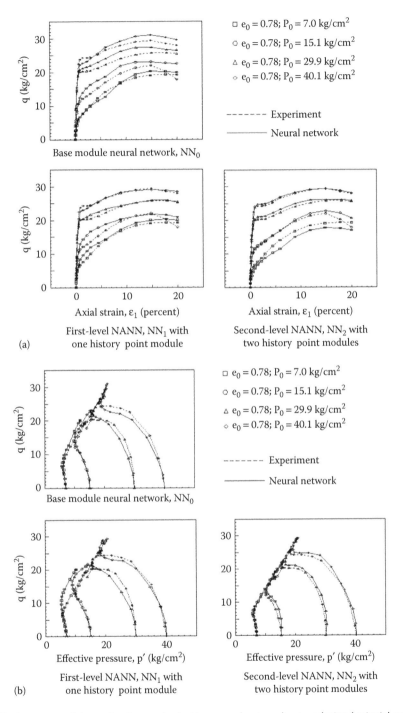

Figure 3.15 Performance of the trained nested adaptive neural network on undrained triaxial compression tests: (a) deviatoric stress q versus the axial strain, (b) deviatoric stress q versus the effective pressure p'. (From Ghaboussi, J. and Sidarta, D.E., *Int. J. Computer Geotechnics*, 22, 29–51, 1998.)

Similar comparisons for the undrained tests are shown in Figure 3.15. Although the base module is able to model the material behavior reasonably well, the addition of one and two history point modules significantly improves the material model. Significant improvements by the addition of the second history point module on the modeling of the undrained behavior of sand can be clearly seen in Figure 3.15. Additional results are available in Ghaboussi and Sidarta (1998), Sidarta (2000).

3.7 MODELING OF HYSTERETIC BEHAVIOR OF MATERIALS

In the previous section, we showed that the NANN was capable of modeling the hysteretic behavior of materials. In this section, we will present another neural network topology that is also capable of modeling the hysteretic behavior of materials (Yun et al., 2008a). In this method, two new variables are introduced. These variables are shown in the following equation and they are illustrated in Figure 3.16.

$$
\begin{cases}
\xi_n = \sigma_{n-1}\,\varepsilon_{n-1} \\
\Delta\eta_n = \sigma_{n-1}\,\Delta\varepsilon_n
\end{cases}
\tag{3.34}
$$

These two variables provide enough information to the neural network that helps it to learn the hysteretic behavior of the material in loading and unloading. The first variable provides information on the current state in the stress–strain space. The second variable indicates the direction of the loading or unloading. Figure 3.16 is for illustrative purposes; it does not give the whole picture. It has one component of stress and one component of strain. In the actual three-dimensional structural systems, there are six components of stress and six components of strain.

The neural network material model with on history point and the two new variables is as follows:

$$
\sigma_n = \mathrm{NN}\big[\varepsilon_{n-1},\sigma_{n-1},\varepsilon_n,\xi_n,\Delta\eta_n; \qquad \big]
\tag{3.35}
$$

The performance of the neural network is demonstrated on the one-dimensional compression experiment on cylindrical plain concrete as shown in Figure 3.17. NANNs were tested on the same experiments and results were shown in Figures 3.11 and 3.12.

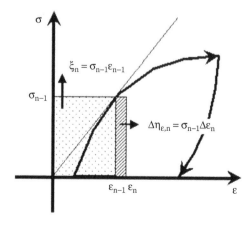

Figure 3.16 The two variables used in the neural network modeling of the hysteretic behavior of materials. (From Yun, G.J. et al., Int. J. Numer. Meth. Eng., 73, 447–469, 2008a.)

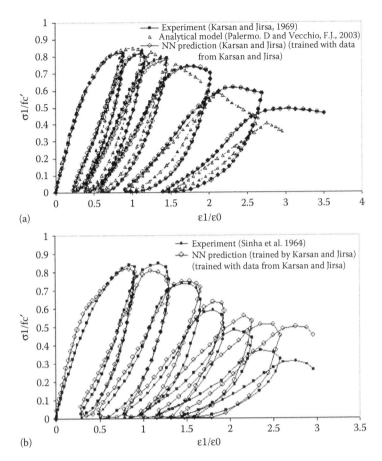

Figure 3.17 Performance of the hysteretic NN in one-dimensional compression experiment on plain concrete sample: (a) training of hysteretic NN, (b) testing of the hysteretic NN. (From Yun, G.J. et al., *Int. J. Numer. Meth. Eng.*, 73, 447–469, 2008a.)

The performance of the neural network material model with two new variables, shown in Figure 3.17, indicates that it is able to model the hysteretic behavior of material with reasonable accuracy, as well as NANN. This neural network was used in modeling of beam-column connections, and the results are presented in Chapter 5, Section 5.10.

3.8 ACQUISITION OF TRAINING DATA FOR NEURAL NETWORK MATERIAL MODELS

In the introduction in this chapter, we discussed the fact that mathematically based material modeling is based on the observed behavior in material tests, fundamental laws mechanics, such as conservations laws, and some assumptions. This has led to the development of fields of material modeling such as elasticity and plasticity that are commonly used in computational mechanics. We have covered this subject in detail in the book (Ghaboussi et al., 2017). When a material model is needed in a finite-element analysis, normally a mathematical model that satisfies the conservation laws and some assumptions is chosen and material tests are used to determine the parameters of the mathematical model. This places some restrictions on material tests. Since the material model represents the behavior of a generic material point, the test specimen, or a sub-region within it, also must represent a material point as closely as possible. This means that the state of stresses and strains within the specimen, or a subregion within it, must be as uniform as possible.

The situation with developing training data for neural network material models is different. The results of uniform material tests intended for determining the parameters of mathematical material models are not sufficient for neural network training. All the points in a uniform material test follow one specific stress path. Therefore, the results of that experiment only contain information about the material behavior along that stress path.

A trained neural network material model that can be used in computational mechanics needs to have comprehensive knowledge of the material behavior in the region of interest in the stress space. This means that it should be trained on a comprehensive training data that contain the information about material behavior along many stress paths in the region of interest in stress space. Since uniform material tests generally follow just one specific stress path, to generate comprehensive training data, we may need to perform many uniform material tests, each subjected to a different stress path. This may not be practical in most cases.

In the early stages of development of neural network material models, the results of uniform material tests were used to explore and evaluate the capability of neural networks to learn constitutive behavior of material. The two examples of uniaxial cyclic behavior of concrete and behavior of sand in drained and undrained triaxial tests fall into this category. These were performed to explore the capability of neural networks, specially NANNs, to learn the complex material behavior. However, the trained neural networks in these examples cannot be used in finite-element simulations involving two- or three-dimensional states of stress.

Material tests are not the only source of information on the material behavior. There are many other potential sources of data that contain information on material behavior. The measured response of a system subjected to a known excitation contains information on the constitutive behavior of the material (or materials) in that system. An example is a structural test. The measured displacements of a structural system that is subjected to known forces contain information on the constitutive properties of the materials within that structural system. This is a more complex inverse problem than the material tests; the forces and displacement are the known input and output of the system and the constitutive properties of the materials within the structural system need to be determined. Unlike material tests, which ideally induce uniform states of stress within the sample, structural tests induce nonuniform states of stresses and strains within the sample. Since points in the specimen follow different stress paths, a single structural test potentially has far more information on the material behavior than a material test in which all the points follow the same stress path.

Everything that a neural network material model learns is necessarily contained in the training data. Conservation laws are satisfied since the training data originates from the response of real materials that presumably do satisfy them. We can also rotate the reference axes, thereby generating additional training data, to ensure that the neural network learns about the frame indifference. However, we make no assumptions about the material behavior itself; the neural network must learn all aspects of the material behavior from the training data. That is why the training data have to be comprehensive. To generate that type of comprehensive data, the alternative to performing many, many uniform material tests subjected to a variety of different stress paths, is to perform *nonuniform* material tests that are similar to structural tests. The state of stress in the sample has to be spatially as nonuniform as possible. This is directly opposite to the conventional thinking about material tests.

In a nonuniform material test, different points in the sample follow different stress paths, thus generating information about material behavior along many, but not all the possible, stress paths within the region of interest in the stress space. Then, the question become how can we extract that information from the results of a nonuniform test to train a neural network? Extracting the information on the material properties from nonuniform structural tests is an extremely difficult problem with the conventional mathematically based methods. This is probably the reason why it has not been attempted in the past. On the other hand, soft computing methods are ideally suited for this type of difficult inverse problem. The learning capabilities of a neural network offer a solution to this problem.

The only way to extract information from the results of a nonuniform test is to simulate that test in a finite-element analysis. But the dilemma is that to do the finite-element simulation we need a trained neural network material model. So, we need a trained neural network material model to perform the finite-element simulation to generate the training data needed to train the neural network material model. This does not appear to be logical. However, there is a way to accomplish this; we can iteratively combine the two tasks of finite-element simulation to generate the data and to use that data to train the neural network material model to be used in the finite-element simulation. This is done by means of the *Autoprogressive Algorithm* that will be described in detail in Chapter 5. Autoprogressive Algorithm is a method for training a neural network to learn the constitutive properties of materials from nonuniform structural tests, and it was first reported in the following reference (Ghaboussi et al.,1998).

But, first we need to discuss how neural network material models can be used as constitutive models directly in finite-element analysis.

3.9 NONLINEAR FINITE-ELEMENT ANALYSIS WITH NEURAL NETWORKS CONSTITUTIVE MODELS

Neural network material models can be used to represent nonlinear material behavior in finite-element analysis, similar to conventional mathematical material models. Nonlinear finite-element analyses are normally performed incrementally, and within each load increment a number of equilibrium iterations are performed to attain satisfactory convergence; more details can be found in the books (Ghaboussi et al., 2017; Ghaboussi and Wu, 2016).

During each iteration, a number of passes through a material model package are needed. A pass through a material model package is needed for each integration point within each element. The algorithm defining a mathematical material model is encoded within the material model package. The input to the material model package includes the current strain increments and the output is the current stresses and, if needed, the tangent stress–strain matrix (the Jacobian). A trained neural network material model can also perform both tasks. Computation of the stress increments from strain increments involve forward passes through the trained neural network material model. There are two methods of determining the tangent stress–strain matrix from the neural network material model that will be discussed later in this section. But first a brief review of finite-element analysis.

The finite-element equations of equilibrium can be written at load step n, in terms of the load vector, **P**, and the internal resisting force vector, **I**.

$$I_n - P_n = 0 \tag{3.36}$$

$$\begin{cases} I_n = I_{n-1} + \Delta I_n \\ P_n = P_{n-1} + \Delta P_n \end{cases} \tag{3.37}$$

The subscripts are the increment numbers.

During each load increment, Equation 3.36 is solved, often using modified Newton–Raphson iterations, which require solution of a system of equations. This task can be performed using either an explicit method such as the preconditioned conjugate gradient method or a direct solution method such as Gauss elimination or Cholesky decomposition. The distinction between these methods is important when a neural network is being used as the material model. The direct solution methods require the formation of the tangent stiffness matrix, whereas the conjugate gradient method does not. These methods are discussed in detail in the book of Ghaboussi and Wu (2016).

Given the strain increments and any other input required by the neural network material model, a single forward pass determines the stress increments. This operation is all that is needed for nonlinear finite-element analysis, if an iterative method such as preconditioned conjugate gradient method is used for the solution of the equilibrium equations. Preconditioned conjugate gradient method requires only the computation of the internal resisting force vector, which can be computed by directly assembling the element contributions:

$$\Delta \varepsilon_n = \mathbf{B} \, \Delta \mathbf{U}_n \tag{3.38}$$

$$\mathbf{I}_n = \mathbf{I}_{n-1} + \sum \int \mathbf{B}^T \Delta \sigma_n \mathbf{NN} \left[\Delta \varepsilon_n, \, , : \, \right] dv \tag{3.39}$$

In this equation, $\Delta \mathbf{U}$ is the incremental displacement vector, and $\mathbf{B}$ is the element strain–displacement matrix. The integration is over the volume of element, and it is performed numerically using Gaussian Quadrature. The summation sign denotes direct stiffness assembly of all elements contributions.

When the tangent stiffness matrix is available, then the incremental finite-element equations can be rewritten as follows:

$$\mathbf{I}_n = \mathbf{I}_{n-1} + \mathbf{K}_t \, \Delta \mathbf{U}_n \tag{3.40}$$

$$\mathbf{K}_t = \sum \int \mathbf{B} \, \mathbf{C}_t \, \mathbf{B} \, dv \tag{3.41}$$

$$\Delta \sigma_n = \mathbf{C}_t \, \Delta \varepsilon_n \tag{3.42}$$

In this equation $\mathbf{K}_t$ is the tangent stiffness matrix, and $\mathbf{C}_t$ is the tangent stress–strain matrix. This leads to the following form of the equilibrium equation that requires solution of system of equations to determine the displacement increment vector.

$$\mathbf{K}_t \, \Delta \mathbf{U}_n = \mathbf{P}_n - \mathbf{I}_{n-1} \tag{3.43}$$

The incremental stress–strain matrix $\mathbf{C}_t$ is needed to update the tangent stiffness matrix. The tangent stiffness matrix is not directly available from the output of the neural network material model. However, they can be determined in two ways: (1) by probing the neural network material model, or (2) directly computing it from the connection weights of the neural network.

We will describe the first method that uses the probing of the neural network. Each probe requires a forward pass through the neural network. For example, in a two-dimensional plane stress or plane strain problem, using a neural network material model with or without history points, three forward passes are needed to determine the three columns of the tangent stress–strain matrix, as shown in the following equation:

$$\begin{cases} C_{ij} = \dfrac{1}{\alpha} \mathbf{C}_j \ \mathbf{NN}\left[\{\Delta\varepsilon_i\}_j, \text{History points:} \quad \right] \quad j = 1, \ 2, \ 3 \\[2mm] \{\Delta\varepsilon_i\}_j = \{\alpha\delta_{ij}\} \end{cases} \tag{3.44}$$

Equation 3.44 describes three probes for j = 1, 2, 3 along the three components of strain, that is, along strain paths $[\alpha, 0, 0]^T$, $[0, \alpha, 0]^T$, and $[0, 0, \alpha]^T$. Each probe generates a stress vector $\mathbf{C}_j$ that, when normalized by α of the jth strain probe, constitutes the jth column of the tangent stress–strain matrix.

It is important to note that the determination of the incremental stress–strain matrix from Equation 3.44 using directional probes will not necessarily produce a symmetric matrix. Trained neural network material models acquire all their knowledge from training data; if knowledge of symmetry of the tangent stress–strain matrix is not implicitly present in the training data, then the neural network will not learn to produce a symmetric constitutive matrix. In addition, the training data are likely to contain enough noise and scatter, so that the stress–strain matrix from directional probes will not be precisely symmetric. If symmetry of the tangent stress–strain matrix is important in the solution algorithm for computational reasons, then it can be imposed; for example, by forming 1/2 $(\mathbf{C} + \mathbf{C}^T)$.

In the second approach, the tangent stress–strain matrix is directly determined from the neural network material model. Tangent stress–strain matrix can be considered the Jacobian of the neural network material model. We will first present the methodology for determining the Jacobian of a general neural network that was first introduced in Hashash et al. (2004) and then specialize it for material model neural networks.

We will consider a general trained neural network with NL layers that include the hidden layers plus the output layer. The activation, or the output, S_i^k of the node i in layer k is given in the following equation:

$$\begin{cases} z_i^k = \displaystyle\sum_{j=1}^{N_{k-1}} w_{ij}^k S_j^{k-1} \\[3mm] S_i^k = f\left(z_i^k\right) \end{cases} \tag{3.45}$$

f is the activation function, w_{ij}^k is the connection weight for the connection from node j in layer k–1 to the node i in layer k, and, N_{k-1} is the number of nodes in the layer k–1. Jacobian of the neural network is given in the following equation:

$$J_{ij} = \frac{\partial\left(S_i^{NL}\right)}{\partial\left(S_j^0\right)} \tag{3.46}$$

$$\begin{cases} S_i^{NL} = \text{output} \\[2mm] S_j^0 \ = \text{input} \end{cases}$$

This equation can be expressed in derivative from one layer to lower layer, starting from the output layer and ending at the input layer:

$$J_{ij} = \frac{\partial\left(S_i^{NL}\right)}{\partial\left(S_k^{NL-1}\right)} \frac{\partial\left(S_k^{NL-1}\right)}{\partial\left(S_m^{NL-2}\right)} \cdots \frac{\partial\left(S_r^1\right)}{\partial\left(S_j^0\right)} \tag{3.47}$$

The terms of this equation can be determined from Equation 3.45, as follows:

$$\frac{\partial\left(S_i^k\right)}{\partial\left(S_j^{k-1}\right)} = \frac{\partial f\left(z_i^k\right)}{\partial\left(z_m^k\right)}\frac{\partial\left(z_m^k\right)}{\partial\left(S_j^{k-1}\right)} = \frac{\partial f\left(z_i^k\right)}{\partial\left(z_m^k\right)} w_{mj}^k \tag{3.48}$$

This equation indicates that the Jacobian of the neural network can be determined from the derivatives of the activation functions at the nodes and the connection weights.

In neural network material models, the input and output are the scaled strains and stresses. Most often the activation function is sigmoid function with constant λ.

$$\begin{cases} \varepsilon_i^{NN} = \dfrac{\varepsilon_i}{s^\varepsilon} & ;\quad 0 \le \varepsilon_i^{NN} \le 1 \\[3mm] \sigma_i^{NN} = \dfrac{\sigma_i}{s^\sigma} & ;\quad 0 \le \sigma_i^{NN} \le 1 \end{cases} \tag{3.49}$$

$$S_i^k = f\left(z_i^k\right) = \frac{1}{1 - e^{-\lambda z}} \tag{3.50}$$

In a three-layer neural network material model (input layer, 2 hidden layers and output layer), the tangent stress–strain matrix is given by the following equation:

$$\begin{aligned} \frac{\partial \sigma_i}{\partial \varepsilon_j} &= \frac{s^\sigma}{s^\varepsilon}\frac{\partial \sigma_i^{NN}}{\partial \varepsilon_j^{NN}} \\[3mm] &= \frac{s^\sigma}{s^\varepsilon}\frac{\partial\left(S_i^3\right)}{\partial\left(S_k^2\right)}\frac{\partial\left(S_k^2\right)}{\partial\left(S_m^1\right)}\frac{\partial\left(S_m^1\right)}{\partial\left(S_j^0\right)} \\[3mm] &= \frac{s^\sigma}{s^\varepsilon}\lambda^3\left\{\left[S_i^3\left(1-S_i^3\right)\right]w_{ik}^3\right\}\left\{\left[S_k^2\left(1-S_k^2\right)\right]w_{km}^2\right\}\left\{\left[S_m^1\left(1-S_m^1\right)\right]w_{mj}^1\right\} \end{aligned} \tag{3.51}$$

This equation is valid for both single module neural networks and neural network material models with history points in NANNs or with other inputs besides the strains at the input layer. The connection weights in the equation are from the base module and the summation is done over the nodes of hidden layers. Then, the question is: how do the history points affect the material properties represented in the tangent stress–strain matrix? This effect is represented through the nodal activations; in computing the nodal activations the input from history point connections must also be included.

The resulting tangent stiffness matrix from this method may also be slightly nonsymmetric, similar to the first method described earlier by probing the neural network. This is due to the fact that neural networks learn from the data with some degree of error tolerance. The example given in reference (Hashash et al., 2004) demonstrates this point. A neural network was trained on the linear elastic material behavior in two-dimensional plane strain case, with elastic modulus E = 500 and Poisson's ratio υ = 0.3. The actual stress–strain relation is

$$\begin{Bmatrix} \sigma_{11} \\ \sigma_{22} \\ \sigma_{12} \end{Bmatrix} = \frac{E}{(1+\upsilon)(1-2\upsilon)} \begin{bmatrix} 1-\upsilon & \upsilon & 0 \\ \upsilon & 1-\upsilon & 0 \\ 0 & 0 & \frac{1-2\upsilon}{2} \end{bmatrix} \begin{Bmatrix} \varepsilon_{11} \\ \varepsilon_{22} \\ \varepsilon_{12} \end{Bmatrix}$$

(3.52)

$$= \begin{bmatrix} 673.1 & 288.5 & 0 \\ 288.5 & 673.1 & 0 \\ 0 & 0 & 192.3 \end{bmatrix} \begin{Bmatrix} \varepsilon_{11} \\ \varepsilon_{22} \\ \varepsilon_{12} \end{Bmatrix}$$

The stress–strain matrix determined from the trained neural network is

$$\begin{Bmatrix} \sigma_{11} \\ \sigma_{22} \\ \sigma_{12} \end{Bmatrix} = \begin{bmatrix} 701.0 & 304.1 & -0.3 \\ 304.7 & 701.1 & -0.1 \\ 0.2 & 0.1 & 193.4 \end{bmatrix} \begin{Bmatrix} \varepsilon_{11} \\ \varepsilon_{22} \\ \varepsilon_{12} \end{Bmatrix}$$

(3.53)

As expected, the neural network stress–strain matrix is not precise. However, it is reasonably close to the exact matrix for most practical applications. This type of minor error in neural network models has been discussed in detail in earlier chapters. Since symmetry is important in most finite element analyses, a symmetric version of the stress–strain matrices can be easily determined:

$$C = \frac{1}{2}\left(C_{NN} + C_{NN}^T\right)$$

(3.54)

$$C = \frac{1}{2}\left(\begin{bmatrix} 701.0 & 304.1 & -0.3 \\ 304.7 & 701.1 & -0.1 \\ 0.2 & 0.1 & 193.4 \end{bmatrix} + \begin{bmatrix} 701.0 & 304.7 & 0.2 \\ 304.1 & 701.1 & 0.1 \\ -0.3 & -0.1 & 193.4 \end{bmatrix}\right)$$

(3.55)

$$= \begin{bmatrix} 701.0 & 304.4 & -0.05 \\ 304.4 & 701.1 & 0 \\ -0.05 & 0 & 193.4 \end{bmatrix}$$

3.10 TRANSITION FROM MATHEMATICAL MODELS TO INFORMATION CONTAINED IN DATA

In working with neural networks, it is always important to remember that the information that the neural network learns is contained in the training data. The principle of functional nonuniqueness, discussed earlier in Chapter 1, means that different neural networks can be trained to learn the information contained in the same data with some degree of imprecision tolerance. In many applications, the data change and the information contained in the data also change and the neural network can be updated with the new data. This is precisely what happens in the autoprogressive algorithm that will be discussed in Chapter 5. At any point in the evolution of the data, it is possible to discard the old trained neural network and start with a new neural network and train it on the expanded and modified dataset.

This observation leads to an important and profound point about the use of neural networks in modeling of physical phenomena such as modeling constitutive behavior of materials. In the mathematically based approaches, we use mathematical functions to model the physical phenomena. The information about the response of the physical system to the external stimuli is contained in that mathematical model.

When we use neural networks to model the physical phenomena, the information about the response of the physical system to the external stimuli is learned by the neural network from its training data. So, if we start thinking in terms of information—rather than mathematical models—then the neural networks are simply tools for extracting the information from the training data and storing that information in its connection weights. It is the learning capabilities of the neural networks that allow extracting the information. Since we can extract the information from data at many stages, and with many different neural networks, it is the data and the information contained in that data that is the soft computing equivalent of the mathematical model and the information contained in the mathematical model. In moving from mathematical modeling to soft computing with neural networks, it is important that we start transitioning our thinking from mathematical models to the information contained in data.

Chapter 4

Inverse problems in engineering

4.I FORWARD AND INVERSE PROBLEMS

The forward and inverse problems are usually considered in the context of input and output to a system or a process, or in the sense of cause and effect, or stimulus and response. There are three basic components and a number of factors involved in the definition of the forward and inverse problems. We will first consider the three basic components: the system dynamics, the input to the system, $\mathbf{X}$, and the output of the system, $\mathbf{Y}$. This relationship is symbolically shown in Figure 4.1.

The input and output of the system are shown in bold letters to signify that they may be vector quantities. The may also be time dependent.

At this point, we are only considering the model of the system, and we are also assuming that the input and output of the system are measurable. We will further assume that two of the three components shown in Figure 4.1 are known, and the third component needs to be determined. This gives rise to three types of problems.

In the first type of problem, the system dynamics and the input to the system are known, and the output of the system needs to be determined. This is the forward problem. It is the most common problem in engineering analysis and computational simulation. We have the equations, or the computational model, describing the behavior of the system to any given input. The results of the engineering analysis, whether it is a simple calculation or highly complex computation, yield the output of the system.

There are two types of inverse problems. In the first inverse problem, the equations of the system dynamic, or its computational model, are known and the output of the system is measured. The objective is to determine the input of the system that caused the measured output. The second inverse problem is when the input and the output are known, and the equations of the system dynamics, or its computational model, need to be determined.

4.2 INVERSE PROBLEMS IN ENGINEERING

We will first consider a very simple problem to outline various aspects of the problem. Consider the example of the system being the discrete model of linearly elastic structural system with a number of degrees of freedom. This could be a finite-element model of the structural system. The input to the system is the load vector $\mathbf{p}$ containing the applied forces and moments at the structural degrees of freedom. The output of the system is the vector $\mathbf{u}$ of the displacements and rotations at the same degrees of freedom. In this case, the system is characterized by the structural stiffness matrix, $\mathbf{K}$, which relates the displacement vector to the load vector, as shown in the following equation:

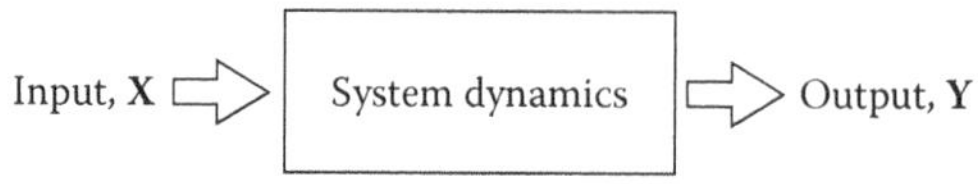

Figure 4.1 Representation of a system, with input and output.

$$Ku = p \tag{4.1}$$

In its most common form, the stiffness matrix is square, and the load vector and the displacement vector are of the same size. Furthermore, if the structure has a sufficient number of supports and no unstable parts, then the stiffness matrix is also nonsingular and can be inverted.

For this example of the structural problem, the forward problem is defined to determine the displacements for the known values of loads. Under the very special circumstances described above, this is routine, and it involves solving the linear system of equations in Equation 4.1. The first inverse problem is to determine the loads that produce known values of displacements. This is even easier to solve, and it involves matrix multiplication of the stiffness matrix and the displacement vector.

The second inverse problem involves determining the structural stiffness matrix from the known loads and the known displacements they produce. This problem is not so easy to solve. Either it can be attempted to directly determine the terms of the stiffness matrix, or an attempt can be made to determine the information needed to reconstruct the stiffness matrix.

The direct determination of the terms of the stiffness matrix is only possible for a single degree of freedom system. Otherwise, there are not enough independent equations to determine the $N(N+1)/2$ terms of an $N \times N$ symmetric stiffness matrix.

The information to reconstruct the stiffness matrix includes the geometry of the structure, nodal coordinates, member connectivity, and the section properties. This information cannot be determined from the loads and the displacements they produce. A limited version of this problem can be tackled in some special cases. For example, it can be assumed that the geometry of the structure is known, including the spatial coordinates of the nodes and the member connectivities, and only the section properties are unknown. This is an easier inverse problem than assuming that nothing is known about the structure. The difficulty is caused by the fact that there is no unique solution.

To slightly complicate the problem, let us assume that the values of the displacements are only known at a few degrees of freedom. The first inverse problem now is restated as follows. The stiffness matrix for the structure is known, and the values of the displacements are known at a few degrees of freedom. Determine the loads that have caused those displacements. This problem has no longer a unique solution. This can be clearly seen by partitioning the degrees of freedom in Equation 4.1 into those with known displacements, u_α, and those with unknown displacements, u_β. The load vector is similarly partitioned into p_α and p_β.

$$\begin{cases} K_{\alpha\alpha}u_\alpha + K_{\alpha\beta}u_\beta = p_\alpha \\ K_{\beta\alpha}u_\alpha + K_{\beta\beta}u_\beta = p_\beta \end{cases} \tag{4.2}$$

In these equations, the stiffness matrices are the result of partitioning of the structural stiffness matrix to separate the degrees of freedom with known displacements from those with unknown displacements.

$$K = \begin{bmatrix} K_{\alpha\alpha} & K_{\alpha\beta} \\ K_{\beta\alpha} & K_{\beta\beta} \end{bmatrix} \tag{4.3}$$

Equation 4.2 shows that any value of the vector of unknown displacements, $\mathbf{u}_\beta$, results in a set of loads in $\mathbf{p}_\alpha$, and $\mathbf{p}_\beta$ which will produce the known displacements $\mathbf{u}_\alpha$. Therefore, this inverse problem also has an infinite number of solutions.

The example of forward and inverse problem in structural analysis is a very simple problem. Practical inverse problems in engineering are far more complex. Every modeling task is an inverse problem. For some stimulus (input), we have some measurement of the response (output) of the system, and we need to identify the system by developing a model for it. This is clearly the second type of inverse problem that we defined earlier. Often the system is too complex, and we may not know whether the measured quantities are sufficient for developing a model of the system. The question is whether some measured quantities contain enough information to characterize the system and build a model.

Constitutive modeling of the material behavior is also the second type of inverse problem. In the laboratory, we perform experiments by applying stresses (input) and measuring the strains (output). By using this information, we seek to develop models of the material behavior. Engineering design, the most basic engineering task, is an ultimate inverse problem. The design specification and design codes and standards define the operating conditions (input) and impose constraints on the response (output) of the system to be designed. The designer seeks a system that will satisfy these input and output requirements. The designer is solving an inverse problem of the second type.

Condition monitoring and damage detection belong to another class of inverse problems. The output of the system, in response to some input, is measured during the inspection, and the state of the system, including the existence of any damage, needs to be determined. This is one of the few inverse problems that are recognized by most engineers as such. However, it is seldom solved directly as an inverse problem.

Control problems are another important class of inverse problems. In this case, we seek a system (control algorithm and control hardware) in order that the system response (output) to certain inputs satisfies some predefined criteria.

In fact, most engineering problems are inverse problems. However, they are seldom posed as inverse problems, and engineers do not often think of them as inverse problems.

All the inverse problems described earlier share some fundamental property: They do not have unique solutions. Theoretically, an infinite number of solutions are available. However, in practice they have a finite number of admissible solutions.

As we will see later, the engineering methodology, which is based on mathematics, is more suitable for solving the forward problems than the inverse problems. As a consequence, almost all the inverse engineering problems are solved as a series of forward problems. A solution is developed, and its behavior is evaluated by a forward analysis. Next, some modifications are made to the solution and it is reevaluated through a forward analysis, and the process is repeated until a satisfactory solution is arrived at.

We will see later that nature has developed very robust and effective methods for solving very difficult inverse problems. The methodology used by nature is fundamentally different than the engineering methodology we use. We can emulate nature's strategy for solving inverse problems in engineering by using neural networks and genetic algorithm.

4.3 INVERSE PROBLEMS IN NATURE

Consider the problems that animals have to deal with to survive in nature and reproduce. They have to recognize their food, their mating partner, and any threats that they face. They use vision, hearing and sense of smell, and possibly some other sensory inputs, such as vibrations. Recognition is the output of the process. Recognition is one of the most difficult inverse problems, if not the most difficult one. Yet the animals solve this problem very effectively, in real time.

In our daily lives, we also solve many difficult inverse problems. Upon hearing a piece of music, we may recognize the instrument. We receive a phone call and may instantly recognize the voice of the caller. We see a face and may instantly recognize it as belonging to a person we know. We can perform many difficult control tasks with ease. We can drive a car and negotiate it around obstacles. And many decisions that we make are the result of solving an inverse problem.

The forward problem in recognition is associating a voice or a face with a person or associating an image with an object. For example, the forward problem in object recognition would be to be told that this is the image of object A. The inverse problem is to be given an image and being asked to identify it. Similarly, the forward problem is to be introduced to an individual and hearing his or her voice and seeing his or her face. The inverse problem is to see a face and hear a voice and attempt to identify the person.

In the case of control problems, we also have forward and inverse problems. For example, a forward problem is to be presented with a route and sequence of actions we have to take in driving a car to navigate around an obstacle. The inverse problem is to be given the objective of navigating the car around an obstacle and being asked to devise a route and take a sequence of actions to achieve the objective.

These are very difficult inverse problems. Since the solutions of these inverse problems were the prerequisite for survival in the animal world, nature had to evolve effective solutions for these problems. There are important lessons for engineering in the strategies used by nature to solve these problems.

The solution strategies that have evolved had to satisfy the basic needs for survival, while neglecting what is not important for survival. Robustness and real-time response are essential for survival. Precision and universality are not important, as discussed briefly in Chapter 1.

We will first define what is meant by robustness in a recognition. Those who work on developing computational methods for recognition, know how difficult this can be. If a predator is to recognize its prey, the recognition must be robust and must take place independent of the background scene, even if the view of the prey is partially blocked, and independent of what else is in the predator's angle of view. Robust recognition is also independent of the location and some variations in color and shape of the prey.

Similarly, when we recognize someone's face, we can still recognize if it was partially blocked, or the person had a different haircut, or was wearing sunglasses. This is what we mean by robust cognition.

It is also important for the survival of the animals that the recognition take place in real time. Any delay in recognizing its prey can be detrimental to a predator's chances of catching its prey. The computational methods that we develop by using serial computers would take too long to solve even the simplest recognition problem, and as such, they would not be appropriate choices. On the other hand, massively parallel computation in brain can produce robust and real-time recognition. That is probably a good reason to think that brain has evolved as a massively parallel computer, not as a serial one.

It is also important to consider what is not important in nature in solution of the inverse problems. Again, if we consider the problem of recognition, it is easy to see that precision and universality are not important for the survival of the animals in nature.

For example, it is sufficient for the predator to know the approximate location of its prey. Any additional effort spent on locating it more precisely is a wasted effort. The predator does not need to know the precise size, shape, and color of its prey. Similarly, when we navigate around an obstacle, the route that we choose is not the precisely optimum choice from all the possible routes. It is sufficient that it be a "near-optimal" route. The term near optimal is also defined rather imprecisely. Again, the difference between the near optimal and the exact optimal is inconsequential in practice, and we usually do not spend any effort in further refining the near-optimal solution. In short, the nature's solution of the inverse problems is accomplished in an imprecision tolerant manner.

The question of universality of solutions has to be addressed in context of the universality of solutions produced by mathematically based methods. For example, the structural stiffness relation in Equation 4.1 is valid for all the possible values of the displacements and loads, irregardless of the fact that the structural displacements and loads are confined to a practical range of values. Outside that range, the response may be nonlinear, and it may lead to failure and collapse of the structural system. The solutions produced by the mathematical equation of the structural stiffness are precise and universal solutions; it indicates that the structural system will behave linearly for all the possible values of loads and displacements.

The methods that have evolved in nature to solve the inverse problems do not produce universal solutions. For example, a predator does not need to recognize every possible object to survive, only a few species that serve as his prey. Similarly, we only recognize the faces of a few acquaintances, not all the possible faces.

In summary, we have seen that nature has evolved effective methods for solving difficult inverse problems. These methods use the power of the massively parallel computation in the brain to produce imprecision tolerant, real-time and robust solutions that are valid only for a limited range of the variables.

4.4 NEURAL NETWORKS IN FORWARD AND INVERSE PROBLEMS

The engineering methods of analysis are based on mathematics. Mathematics is also considered the language of mechanics and physics. It is used to describe natural phenomena, and it forms the basis of all modeling, analysis, and computational methods. As we have discussed before, modeling and analysis are forward problems. Mathematics is also used to solve inverse problems, wherever possible. We have also briefly discussed that the mathematically based methods are more suitable for forward problems than the inverse problems.

Neural networks are capable of solving most of the engineering inverse problems. We will illustrate this capability by presenting some examples in this and next chapter. One of the reasons that mathematical methods are not suitable for solving engineering problems is that most of the engineering inverse problems do not have unique solutions. It is important to point out that nonuniqueness is mainly the consequence of universality and precision in the mathematical definition of the inverse problems. Lack of universality and imprecision tolerance enables the neural networks to find the most relevant solution in the inverse problems.

We need to further explore the nonuniqueness in type-one inverse problems where the system and the output of the system are known and the input needs to be determined. There are three main reasons for the nonuniqueness.

1. Mathematical representation of the system may have many, often infinite, number solutions in its universal representation. Unique solution may exist from practical point of view within the range of interest.
2. Lack of sufficient information in the output of the system may also lead to the possibility of multiple solutions. If complete information in the input and output of the system are available, often the solution of inverse problem has a unique solution. In these

cases, the forward and the inverse problems are similar and neural networks can learn either routinely. Incomplete output information often leads to nonuniqueness. Neural networks may find a solution in the range of interest. We will present an example of this case in the next section.

3. In some cases, the engineering problem is defined in terms of forward problem, like determining the response spectrum from earthquake accelerograms. The forward problem in these cases loses enough information so that the inverse problem does not have unique solution. For example, an infinite number of earthquake accelerograms will produce the same response spectrum. In Section 4.8 in this chapter we will present the example of developing earthquake accelerograms from the response spectrum.

4.5 ILLUSTRATIVE EXAMPLE

Next, we will use the example of a plane frame structure shown in Figure 4.2 to further explore the role of sufficient and insufficient information in inverse problems and the performance of neural networks.

The lateral stiffnesses of the floors are given below:

$$
\begin{cases}
K_1 = 240,000 \text{ kN/m} \\
K_{2-4} = 220,000 \text{ kN/m} \\
K_{5-7} = 200,000 \text{ kN/m} \\
K_{8-10} = 180,000 \text{ kN/m}
\end{cases}
$$

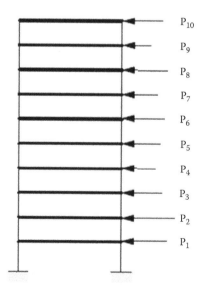

Figure 4.2 Ten-story plane frame structure.

The force–displacement relation through the structural stiffness matrix is as follows:

$$10{,}000 \begin{bmatrix} 24 & -24 & & & & & & & & \\ -24 & 46 & -22 & & & & & & & \\ & -22 & 44 & -22 & & & & & & \\ & & -22 & 44 & -22 & & & & & \\ & & & -22 & 42 & -20 & & & & \\ & & & & -20 & 40 & -20 & & & \\ & & & & & -20 & 40 & -20 & & \\ & & & & & & -20 & 38 & -18 & \\ & & & & & & & -18 & 36 & -18 \\ & & & & & & & & -18 & 18 \end{bmatrix} \begin{Bmatrix} u_1 \\ u_2 \\ u_3 \\ u_4 \\ u_5 \\ u_6 \\ u_7 \\ u_8 \\ u_9 \\ u_{10} \end{Bmatrix} = \begin{Bmatrix} p_1 \\ p_2 \\ p_3 \\ p_4 \\ p_5 \\ p_6 \\ p_7 \\ p_8 \\ p_9 \\ p_{10} \end{Bmatrix}$$

$$(4.4)$$

We can see that there is a unique relationship between the force and displacement vectors. The forward and inverse problems in this case are almost equivalent, and neural networks can learn either. In structural engineering, normally the forward problem is considered when forces are known (input) and displacements are determined (output). In that case, the inverse problem will be when the displacements are known (input) and forces are determined (output). If we provide all the displacements at the ten degrees of freedom as input, then the output is unique and the neural network can learn the inverse problem uniquely, similar to learning the forward problem. We will use this example to first demonstrate how well can the neural network learn the unique inverse problem. Next, we will explore the performance of the neural network when fewer displacements are provided as input and the inverse problem does not have a unique solution.

The neural network in the following equation was trained to learn the inverse problem with all the ten displacements as input:

$$p = NN[u : 10 | 10 | \ 10 | 10] \qquad (4.5)$$

The training data consisted of 100 randomly generated load vectors and their corresponding displacement vectors computed from Equation 4.4. Applied loads at each degree of freedom were in the range of 0–1000. We use the following five load cases to study the performance of the trained neural network.

Test case 1:

$$\begin{cases} P = [10, 100, 200, 300, 400, 500, 600, 700, 800, 900] \\ u = [0.0188, 0.0392, 0.0592, 0.0783, 0.0978, 0.1153, 0.1303, 0.1437, 0.1531, 0.1581] \end{cases}$$

Test case 2:

$$\begin{cases} P = \begin{bmatrix} 500, & 500, & 500, & 500, & 500, & 500, & 500, & 500, & 500, & 500 \end{bmatrix} \\ u = \begin{bmatrix} 0.0208, & 0.0413, & 0.0595, & 0.0754, & 0.0904, & 0.1029, & 0.1129, & 0.1212, & 0.1268, & 0.1295 \end{bmatrix} \end{cases}$$

Test case 3:

$$\begin{cases} P = \begin{bmatrix} 200, & 400, & 600, & 800, & 900, & 900, & 800, & 600, & 400, & 200 \end{bmatrix} \\ u = \begin{bmatrix} 0.0242, & 0.0496, & 0.0733, & 0.0942, & 0.1132, & 0.1277, & 0.1377, & 0.1443, & 0.1477, & 0.1488 \end{bmatrix} \end{cases}$$

Test case 4:

$$\begin{cases} P = \begin{bmatrix} -100, & -100, & -100, & -100, & -100, & -100, & -100, & -100, & -100, & -100 \end{bmatrix} \\ u = \begin{bmatrix} -0.0042, & -0.0083, & -0.0119, & -0.0151, & -0.0181, & -0.0206, & -0.0226, \\ -0.0242, & -0.0253, & -0.0259 \end{bmatrix} \end{cases}$$

Test case 5:

$$\begin{cases} P = \begin{bmatrix} 1000,1000,1000,1000,1000,1000,1000,1000,1000,1000 \end{bmatrix} \\ u = \begin{bmatrix} 0.0417,0.0826,0.1189,0.1508,0.1808,0.2058,0.2258,0.2424,0.2535,0.2591 \end{bmatrix} \end{cases}$$

The load vectors in the first three cases are within the range of the loads used in generating the training data. If the neural network has learned the structural behavior within the range of loads in the training data, it should be able to give reasonable values of the loads in the first three test cases. The results are shown in Figure 4.3.

The neural network response for these three test cases are very close to the actual values of the loads, as was expected since they are within the range of force in the training data. We can see that the performance of the trained neural network for the inverse problem is very similar to a trained neural network for the forward problem. This is expected as there is a unique relationship between the force and displacement vector when complete data are used.

Test case 4 falls outside the range of forces in the training data that were between 0 and 1,000. Therefore, we expect that the trained neural network will not perform well, as can be seen in Figure 4.4.

The forces in the tests case 5 are at the outer limit of the range of the training data for the inverse neural network. We have discussed in Chapter 2 that the performance of the trained neural network deteriorates as we approach the outer limits of the range of the training data. We can clearly see this in Figure 4.5.

The first neural network for the inverse problem of the structural frame was trained with the complete data in the force and displacement vector that results in unique relationship between the input and output. We have seen that the training and the performance of the inverse neural network is similar to a forward problem. The situation is very different when the input information is not complete. We will explore this for the structural frame problem when we provide incomplete information on displacements to the inverse neural network as in the following equation:

$$p = NN\begin{bmatrix} u : 8 \mid 15 \mid & 15 \mid 10 \end{bmatrix} \tag{4.6}$$

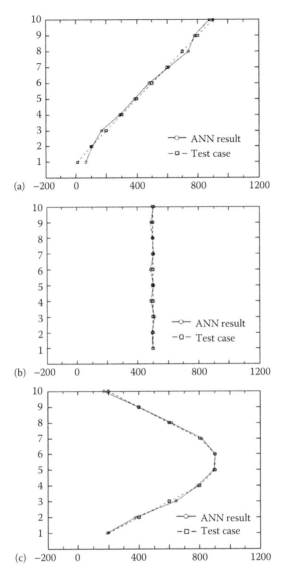

Figure 4.3 Performance of the inverse neural network trained with complete input and output data for the structural frame problem on three cases that are within its training data. (a) Test case 1, (b) Test case 2, and (c) Test case 3.

The output of this neural network has the forces in all the ten degrees of freedom. At the input, we have only provided displacements at 8 of the ten degrees of freedom. The 8 input displacements were at degrees of freedom 1–3, 5–7, 9, and 10. So, we left out the displacements at degrees of freedom 4 and 8. The training data consisted of 100 randomly selected force vectors and their corresponding displacement vectors computed from Equation 4.4. The forces in the training data were in the range of 0 to 1000.

First, we observe that because of nonuniqueness it took longer and more nodes in hidden layers to train this neural network. Neural network in Equation 4.5 had 10 nodes in each hidden layer, whereas the neural network in Equation 4.6 required 15 nodes in each hidden layer.

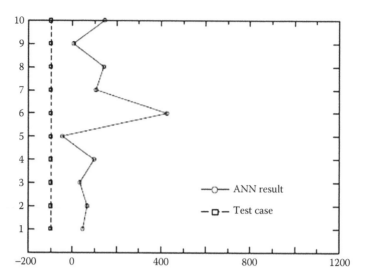

Figure 4.4 Performance of the inverse neural network trained with complete input and output data for the structural frame problem on test case 4 that has loads that fall outside the range of forces used in the training data. As expected the performance of the neural network is not good.

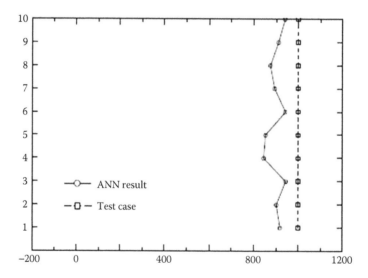

Figure 4.5 Performance of the inverse neural network trained with complete input and output data for the structural frame problem on test case 5 that has loads that are at the outer limit of the range of forces used in the training data. We expect the performance to deteriorate as we approach the limits of the training data.

As there are no unique solutions, it is interesting to explore what this neural network has learned. Again, we explore this by applying the test cases. Figure 4.6 shows the forces generated by the neural network for test cases 1, 2, and 3.

We can see that the output of neural network gives mainly oscillating forces around the actual forces that produced the input to the neural network. We know that because of

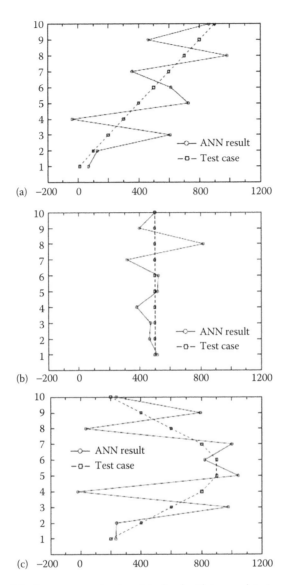

Figure 4.6 Performance of the inverse neural network trained with incomplete input data for the structural frame problem on test cases 1, 2, and 3. (a) Test case 1, (b) Test case 2, and (c) Test case 3.

nonuniqueness, there are many force vectors that can produce the eight input displacements. The question is: Are the force vectors shown in Figure 4.6 one of those force vectors for each of the three test cases. To explore this question, we apply the output force vectors to the frame structure and compute the displacements, shown in Figure 4.7.

It is interesting to see that when we apply the forces from the output of the neural network shown in Figure 4.6 to the frame structure, we compute displacements shown in Figure 4.7 that closely match the actual displacements used as input to the neural network. This means that with the incomplete input, the neural network is able to produce one of the nonunique solutions in the vicinity of the expected solution. This is an important observation; although theoretically there are an infinite number of solutions, neural network is able to solve this inverse problem, based on imprecision tolerance and nonuniversality. Neural network learns

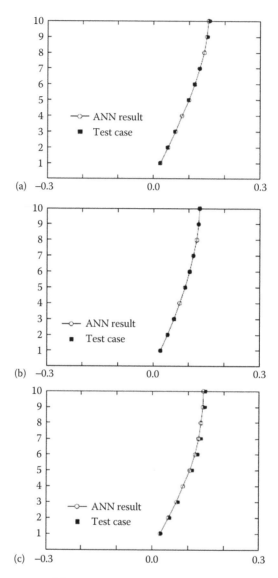

Figure 4.7 Displacements computed by applying the force vector output of the inverse neural network trained with incomplete input data (shown in Figure 4.6) to the structural frame problem for test cases 1, 2 and 3. (a) Test case 1 (b) Test case 2 and (c) Test case 3.

the inverse problem from the training data that only covers small region in the solution space with fewer nonunique solutions.

The neural network model differs from the mathematically based model in fundamental ways. It is closer to modeling that occurs in nature. It is like building a mental image. Unlike mathematical models, it is not precise. It has learned the force–displacement relations from a training dataset in an imprecision tolerant way. It is also not a universal model. It can only represent the force–displacement relations within the range covered in the training set. It is also not unique, in the sense that many similar neural networks, but with different internal architectures, can learn to represent the structural force–displacement relations with a comparable degree of accuracy.

Engineering problems share many of the attributes of the biological problems described in the previous section. Varying degrees of precision may be important in some engineering problems. However, the absolute precision or exactness is not required. Universality is also not an important consideration. Modeling and analysis are useful within the practical ranges of parameters and variables of the problem. Uniqueness of the mathematical representations is also of no consequence in engineering problems.

The major difference between the mathematical models and neural network models is in inverse problems that do not have unique solutions. The mathematical models cannot deal with inverse problems that have nonunique solutions. On the other hand, neural networks can learn these inverse problems as easily as they can learn the forward problems.

The main question is: How does the neural network learn a reasonable and plausible relationship where mathematically unique relationships do not exist? The situation is very similar to the problem of recognition in animals and humans. How do we learn to recognize voices and faces when mathematically unique and universal relationships do not exist? We explore this question in the next section.

4.6 ROLE OF PRECISION, UNIVERSALITY, AND UNIQUENESS

To explore the questions posed at the end of the previous section, we should first consider why the simple mathematical approaches to these inverse problems are usually not successful. The reason can be attributed to basic property of the mathematical methods that operate with absolute precision and produce universal solutions. The lack of unique solutions is the primary obstacle for the mathematically based methods.

On the other hand, the neural network–based methods operate with a certain degree of imprecision tolerance, and they produce only locally admissible, rather than universal, solutions. Consider the problem of recognition of faces. The basic strategy for solving this problem is to learn the information for the inverse problem from a set of forward problems. There are three major elements in this sentence that are the basis for inverse problem solving by both biological systems and neural networks. These three elements are discussed below.

4.6.1 Learning

Learning is at the root of successful solution of inverse problems. We learn how to solve the inverse problems, we learn to recognize faces and voices. The neural network learns the inverse associations. Learning is acquiring some knowledge from a set of examples. Learning, of course, is more than just a simple acquisition of knowledge and filing it away. In that case, we can only retrieve exactly what has been learned. Learning in our brains and in neural networks occurs with some degree of generalization that is used in the retrieval of that information. It is this generalization capability that requires learning with a certain degree of imprecision tolerance.

4.6.2 Learning from forward problems

The information for the inverse problem is learned from the solution of forward problems. In the structural problems, we apply loads and measure displacements at a number of degrees of freedom. We repeat this process for different load cases and develop the data for training of the inverse neural network. The neural network learns the inverse problem from its training data. We do the same in recognizing voices and faces. We learn the inverse problem from the forward problems. The first time we see a face or hear a voice we know the person, and we learn this association.

4.6.3 Learning from a set of forward problems

Finally, the third basic element is that the learning takes place from a "finite set" of forward problems. This is the primary reason why the imprecision-tolerant learning of the inverse problem does not produce universal solutions, as we would expect from mathematically based methods. It produces only locally admissible solutions compatible with the information contained in the training dataset, which comes from a finite set of forward problems. These points will be further explored and demonstrated through some applications to engineering problems from the author's own work.

In summary, we have seen that neural network solutions to the inverse engineering problems are the result of learning from a set of forward problems. Neural network solutions differ from the mathematically based methods mainly in the fact that they are nonuniversal and have a certain degree of imprecision tolerance. The imprecision also implies functional nonuniqueness; more than one neural network can represent the same inverse problem within the same degrees of imprecision.

Next, we will address the question of how neural networks can learn an inverse problem that does not have a unique solution.

4.7 UNIVERSAL AND LOCALLY ADMISSIBLE SOLUTIONS

The input–output relations for the forward problem of the generic system shown in Figure 4.1 can be expressed in the following equation:

$$\mathbf{Y} = \mathbf{f}_{xy}(\mathbf{X}) \tag{4.7}$$

In this equation, $\mathbf{X} = \{x_1,..., x_n\}$ is the n-dimensional input vector, and $\mathbf{Y} = \{y_1,..., y_m\}$ is the m-dimensional output vector. It will be assumed that the output $\mathbf{Y}$ can be uniquely determined from the input $\mathbf{X}$. The inverse problem is expressed in the following equation:

$$\mathbf{X} = \mathbf{f}_{yx}(\mathbf{Y}, \Lambda) \tag{4.8}$$

In this equation, $\Lambda = \{\lambda_1,..., \lambda_k\}$ is the k-dimensional vector representing either the missing information or the information that gets lost in the forward problems.

An example of the missing information is the structural problem described earlier. Recall that the forces were the input and the displacements were the output. In the incomplete version of the inverse problem, it was assumed that the displacements were measured at a few degrees of freedom. The unmeasured displacements constitute the missing information.

In the next section, we will present a case study of generating spectrum compatible artificial earthquake accelerogram. This is an example of a case where information gets lost in the forward problem of computing the response spectrum from an accelerogram. As a consequence, the inverse problem does not have a unique solution.

The inverse function $\mathbf{f}_{yx}$ may be single-valued function and have a unique solution, or it may be a multiple-valued function and the solution may be nonunique. As an example of the latter, consider the sin function as the system equation.

$$Y = A \sin(X) \tag{4.9}$$

Clearly, this is a single-valued function. However, its inverse problem, expressed in the following equation, is not.

$$X = \frac{1}{A} \text{ arc } \sin(Y) \tag{4.10}$$

Clearly, this function has an infinite number of solutions.

In all the three examples described above, incomplete structural measurements, spectrum compatible accelerograms, and arcsin function, neural networks are able to find acceptable solutions. Unique universal solutions with mathematical precision do not exit. However, locally admissible solutions within a degree of imprecision do exist, and neural networks find and learn these solutions.

The difference between the universal solutions and locally admissible solutions can best be demonstrated by the example of arcsin function. The training cases have been generated for the value of x in the interval [a, b]. For each value of X, the corresponding value of Y is determined from the solution of the forward problem. The inverse neural network is then trained with these data.

$$X = NN[Y : 1 \mid 3 \mid 3 \mid 1] \tag{4.11}$$

The performance of the neural network is shown in Figure 4.8. It is clear that the neural network has not learned the arcsin function. It has only learned to approximate it within the range [a, b]. This approximate is a single-valued function.

In solving the inverse problem, the neural network learns to approximate the actual multivalued inverse relation with a single-valued function over the range covered by the training data.

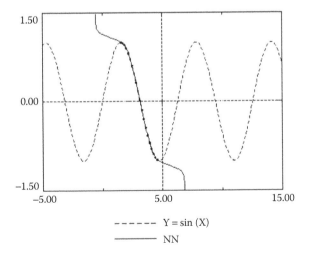

Figure 4.8 Neural network has learned the inverse problem of arcsin (x).

4.8 INVERSE PROBLEM OF GENERATING ARTIFICIAL EARTHQUAKE ACCELEROGRAMS

4.8.1 Preliminaries

Earthquake ground accelerograms are used in the seismic design of civil structures to withstand the effects of earthquakes. Some typical ground acceleration time histories (also referred to as accelerograms) recorded during earthquakes are shown in Figure 4.9.

These accelerograms are recorded at a specific site at regular time intervals, usually at 0.02-second intervals. At each point, recordings are made in three directions: two horizontal directions and one vertical direction. The horizontal ground accelerations are of more interest, since they have the potential for causing significant damage.

Ground acceleration time histories vary as a function of the earthquake source, the magnitude of the earthquake, the distance from source to site, and the ground conditions at the site. Wherever actual recorded earthquakes are available, they are used in evaluating the seismic response of the structures and in seismic design. In some cases, there are no recorded earthquake ground accelerations that match a specific source, distance and site condition. In those situations, artificially generated earthquakes ground accelerations may be used. There has been considerable amount of research on this subject over the past several decades, and it is still an important area of research. We will present an application of neural networks in generating artificial earthquake ground acceleration, as an example of an inverse problem in engineering. This example is from the work of the author and his former graduate student, (Lin, 1999; Ghaboussi and Lin, 1998; Lin and Ghaboussi, 2001).

In addition to accelerograms, their spectra are also used in seismic design. There are many ways of defining the spectra. The two important types are Fourier spectra and response spectra.

The Fourier spectrum is defined by the following equation, and it has real and imaginary parts.

$$A(\omega) = A_r(\omega) + iA_i(\omega) = \int_0^\infty \exp(-i\omega t)\, \ddot{x}_g(t)\, dt \qquad (4.12)$$

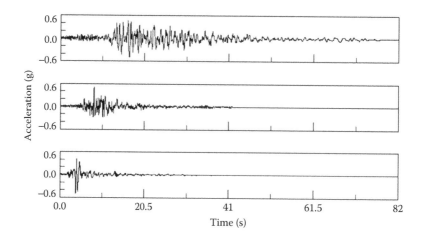

Figure 4.9 Typical recorded earthquake ground accelerograms.

where $\ddot{x}_g(t)$ is the earthquake ground acceleration, and ω is the frequency. One of the important properties of the Fourier transform is that it is reversible. This implies that in computing the Fourier spectrum all the information content of the accelerogram is preserved, so that the accelerogram can be computed from its Fourier spectrum by inverse Fourier transform.

$$\ddot{x}_g(t) = \int_{-\infty}^{\infty} \exp(i\omega t)\, A(\omega)\ dt \qquad (4.13)$$

Structural engineers prefer to use *response spectrum* in evaluating the seismic response of structures and in designing structures against the destructive effects of earthquakes. A response spectrum is defined as the maximum response of an idealized damped single degree of freedom structural system subjected to the earthquake accelerogram, $\ddot{x}_g(t)$, as the ground motion. The response, $x(t)$, of the single degree of freedom structural system is determined from the following differential equation of motion.

$$\ddot{x}(t) + 2\xi\omega\dot{x}(t) + \omega^2 x(t) = -\ddot{x}_g(t) \qquad (4.14)$$

In this equation, ω is the frequency of the single degree of freedom system, and ξ is its damping ratio. The pseudo velocity response spectrum, $S_v(\omega, \xi)$, is defined by the following equation (Newmark and Hall, 1982):

$$S_v(\omega, \xi) = \omega \max|x(t)| \qquad (4.15)$$

The pseudo velocity response spectrum, $S_v(\omega, \xi)$, is normally plotted as a function of the frequency, ω, for different values of the damping ratio, ξ, of the single degree of freedom structural system.

The design codes and design specifications for important facilities may require the use of the "design response spectra." A pseudo velocity response spectrum is determined from a single accelerogram, whereas the design response spectra are determined from a group of recorded or synthesized accelerograms through some predetermined statistical operation, such as a simple average, mean plus one sigma or the envelope.

Unlike the Fourier spectra, the pseudo velocity response spectra are irreversible. Some of the information content of the accelerogram is lost in computing the response spectrum of an accelerogram. Consequently, the accelerogram cannot be uniquely determined from a response spectrum. This is precisely the problem that we are addressing here.

4.8.2 Problem definition

Given a design response spectrum, generate one or more artificial accelerograms that are compatible with design response spectrum, that is, the response spectra of the artificially generated accelerograms must closely approximate the design response spectrum.

The system and its input and output for this problem are shown in Figure 4.10. The system in this case is characterized by Equations 4.14 and 4.15. The input to the system is the accelerogram, and the output of the system is the pseudo velocity response spectrum. Then the forward problem is computing the pseudo velocity response spectrum from the accelerogram, and the inverse problem is determining an accelerogram from a response spectrum. Similar to the vast majority of inverse problems, this one also does not have a unique solution. Infinite number of time histories may have a response spectrum that closely approximates the given design spectrum. However, many of them may not have the appearance of

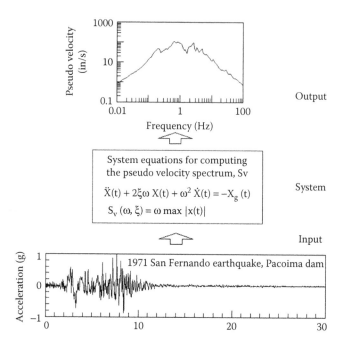

Figure 4.10 The input, output and the system in the forward problem of determining the response spectrum of an earthquake accelerogram.

the earthquake accelerograms. Admissible solutions are those time histories that have the basic characteristics of an earthquake accelerograms.

4.8.3 Neural network approach

The outline of the method for generating multiple artificial earthquake accelerograms is shown in Figure 4.11. Intention is to train a neural network or a series of neural networks to relate the discretized vector of the pseudo velocity response spectrum to the discretized real and imaginary parts of the Fourier spectrum of the artificial accelerogram. The artificial accelerogram is then computed by inverse Fourier transform. The input can either be an actual pseudo velocity spectrum of an accelerogram, like the one in Figure 4.10, or a design response spectrum, like the one shown in Figure 4.11. Of course, the difference is that the design response spectrum is not the actual pseudo velocity spectrum of an accelerogram, but it is determined from a group of pseudo velocity spectra through some statistical method, and it has also been smoothed. The neural networks shown in Figure 4.11 can also be expressed in the following equation:

$$\left[A_r(\omega),\ A_i(\omega)\right] = NN\left[S_v(\omega);\quad\right] \tag{4.16}$$

These neural networks were trained with the accelerograms recorded during earthquakes along with their response spectra. The recorded accelerograms could be classified into different groups with common characteristics such as duration, distance from source, source characteristics, and site conditions. Each group of accelerograms could then be used to train a specific neural network.

Accelerograms, or their Fourier Transforms, carry some essential information needed to reconstruct the accelerogram and some nonessential information and noise. The neural

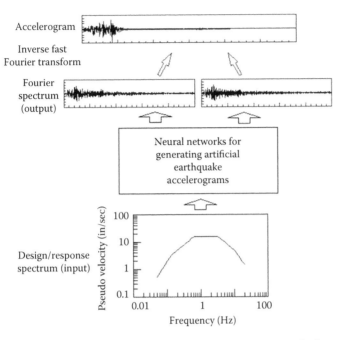

Accelerogram

Inverse fast
Fourier transform

Fourier
spectrum
(output)

Neural networks for
generating artificial
earthquake
accelerograms

Design/response
spectrum (input)

Figure 4.11 Schematics of the neural network based method for generating multiple artificial earthquake accelerograms.

network shown in Figure 4.11 must learn to associate the response spectra to the essential information carried by the accelerograms. The two tasks of association and sorting out the nonessential information were separated into two separate neural networks.

First, the recorded accelerograms were used to train a replicator neural network. The input and output of the replicator neural network were the real and imaginary parts of the Fourier transform of the accelerograms. One of the replicator neural networks used in this study is shown in Figure 4.12. Three replicator neural networks were trained for long, medium, and short duration accelerograms. The technical definition of the duration is given in Lin and Ghaboussi (2001). These neural networks are expressed in the following equation:

$$\begin{cases} \left(A_r + A_i\right)_{LD} = NN_{LD}\left[\left(A_r + A_i\right)_{LD}; \ 4096 \mid 210 \mid 40 \mid 210 \mid 4096\right] \\ \left(A_r + A_i\right)_{MD} = NN_{MD}\left[\left(A_r + A_i\right)_{MD}; \ 2048 \mid 180 \mid 40 \mid 180 \mid 2048\right] \\ \left(A_r + A_i\right)_{SD} = NN_{SD}\left[\left(A_r + A_i\right)_{SD}; \ 1024 \mid 150 \mid 40 \mid 150 \mid 1024\right] \end{cases} \qquad (4.17)$$

The subscripts LD, MD, and SD stand for long, medium, and short duration accelerograms. Since fast Fourier transform was used, the number of input nodes is dependent on the duration. The numbers of input and output nodes are determined by the duration and the use of fast Fourier transform. Each neural network had three hidden layers, as shown in Figure 4.12. The middle hidden layer, with 40 nodes, is the compression layer.

The replicator neural networks perform two functions. The lower part of the neural network, from the input layer to the middle hidden layer (compression layer), is the encoder,

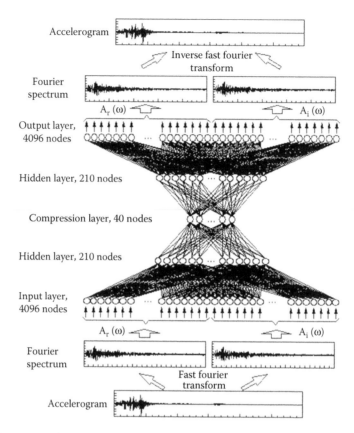

Figure 4.12 Replicator neural network used for data compression of the Fourier spectra. (From Lin, C.-C. J. and Ghaboussi, J., *Int. J. Earthquake Eng. Struct. Dyn.*, 30, 1021–1042, 2001.)

and it performs the task of data compression. The upper part of neural network, from the compression layer to the output layer, does the opposite. It is the decoder and it does the task of data decompression. The activations of the nodes in the middle hidden layer can be thought of as the *compressed Fourier spectrum.*

The upper portions of the trained replicator neural networks are used in this study. They are intended to receive the compressed Fourier transforms as input and give the full Fourier transforms of the artificial accelerograms as output.

Another set of neural networks was needed to relate the response spectrum to the compressed Fourier transforms. To generate multiple accelerograms from a single input response spectrum, these neural networks needed to have a stochastic element. A random temperature-dependent noise term was added to the activation function of the neurons, as shown in the following equations:

$$z_i = \sum_{j=1}^{m} W_{ij} S_j + I_i \qquad (4.18)$$

$$S_i = f(z_i) = \frac{1}{1 + \exp(-\lambda z_i)} + \varepsilon \qquad (4.19)$$

$$S_i = \begin{cases} 0 & \text{if } f(z_i) \leq 0 \\ 1 & \text{if } f(z_i) \geq 1 \\ f(z_i) & \text{Otherwise} \end{cases} \tag{4.20}$$

where $\varepsilon \sim N(0, \tau)$ is the Gaussian noise with zero mean.

Again, three neural networks were trained for the long, medium, and short duration accelerograms. Recorded earthquake accelerograms were used in the training of these neural networks. The trained neural networks are shown in Figure 4.14, and they are expressed in the following equation:

$$\begin{cases} \left(A_{(r+i)c-LD} \right) = NN_{ST-LD} \left[S_v; \ 50 \mid 45 \mid 45 \mid 40 \right] \\ \left(A_{(r+i)c-MD} \right) = NN_{ST-MD} \left[S_v; \ 50 \mid 45 \mid 45 \mid 40 \right] \\ \left(A_{(r+i)c-SD} \right) = NN_{ST-SD} \left[S_v; \ 50 \mid 35 \mid 35 \mid 40 \right] \end{cases} \tag{4.21}$$

Figure 4.13 also shows multiple compressed Fourier spectra generated as the output of each neural network from a single design spectrum used as the input. The compressed

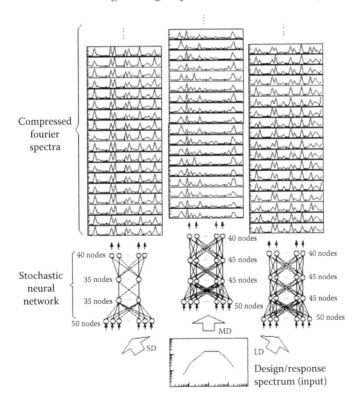

Figure 4.13 The stochastic neural networks for short, medium and long duration accelerograms produce multiple compressed Fourier spectra from a single design response spectrum. (From Lin, C.-C. J. and Ghaboussi, J., *Int. J. Earthquake Eng. Struct. Dyn.*, 30, 1021–1042, 2001.)

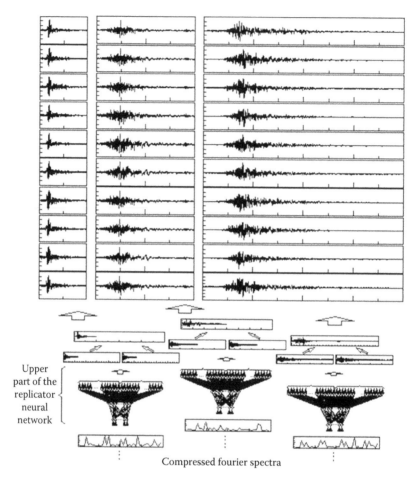

Figure 4.14 Artificial accelerograms, with short, medium and long durations generated from compressed Fourier spectra by the upper half of the replicator neural networks. (From Lin, C.-C. J. and Ghaboussi, J., *Int. J. Earthquake Eng. Struct. Dyn.*, 30, 1021–1042, 2001.)

Fourier spectra are used as input to the upper half of the trained replicator neural networks, expressed in Equation 4.17 and shown in Figure 4.12, to produce multiple artificial earthquake accelerograms, as shown in Figure 4.14.

The whole process of generating multiple artificial earthquake accelerograms from response spectra is illustrated in Figure 4.15. The figure also shows the two-stage neural network system.

4.8.4 Discussion

We have just described the application of neural networks in directly solving an inverse problem that, similar to the majority of inverse problems in engineering, does not have a unique solution. The methodology described in this section is capable of producing admissible solution; the artificial earthquake accelerograms produced by the method have all the characteristics of actual recorded earthquake accelerograms.

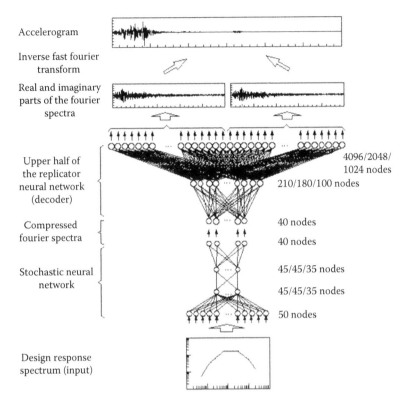

Accelerogram

Inverse fast fourier transform

Real and imaginary parts of the fourier spectra

Upper half of the replicator neural network (decoder) — 4096/2048/1024 nodes / 210/180/100 nodes

Compressed fourier spectra — 40 nodes / 40 nodes

Stochastic neural network — 45/45/35 nodes / 45/45/35 nodes / 50 nodes

Design response spectrum (input)

Figure 4.15 The process of generating multiple artificial earthquake accelerograms from a single response spectrum with the two-stage neural network system is illustrated. (From Lin, C.-C. J. and Ghaboussi, J., *Int. J. Earthquake Eng. Struct. Dyn.*, 30, 1021–1042, 2001.)

It is interesting to note the analogy of how this method operates and how certain inverse problems are solved by animals and human beings in nature. In this method, a series of forward problems are solved to compute the pseudo velocity response spectra for a number recorded earthquake accelerograms. The inverse of the information generated by these forward problems is learned by the neural network. It is as though the neural network builds a library of the information generated by the solution of the forward problems in a way that the retrieval also has a generalization capability.

This is similar to how we learn to recognize voices and faces. We associate each person we know with their voices and faces. These are clearly the solutions to some forward problems. We then learn the information generated by the forward problems in a way that the inverse retrieval is possible. Moreover, we learn that information in a robust manner, far more robust than any neural network is capable of learning.

Again, we observe that both the neural network and the biological systems sacrifice universality, precision, and functional uniqueness to solve a difficult inverse problem. In the same way that we are not capable of recognizing all the voices and faces, the proposed method also does not find universal solutions. It only learns the information contained in the training data set. That is the reason for grouping the accelerograms into sets with common characteristics, such as duration, and training a separate neural network with each group.

4.9 EMULATOR NEURAL NETWORKS AND NEUROCONTROLLERS

In the previous sections in this chapter, we presented two examples of the type-one inverse problems where the output of the system is known, and the input is determined. In this section and the next chapter, we will discuss application of neural networks in type-two inverse problems. We have seen that in type-two inverse problems, the input to the system and its output are known and neural networks are used to learn to represent a model of the system. We will present and discuss an example of soft computing approach in a type-two inverse problem in this section.

The example is the neurocontrol of structures subjected to earthquake ground motion. The experimental setup is shown in Figure 4.16. The structure is a three-story single span steel frame setup on an earthquake shaking table. An actuator on the ground floor is connected to a cable system that applies the control force to the first floor. The objective is for the actuator to apply a force in the first floor to reduce the overall motion of the structure caused by the earthquake ground motion. The control system determines the magnitude to control force.

The system in this case is the combination of structural frame and the actuator and the cable system on the shaking table that is subjected to earthquake ground acceleration $\ddot{X}_g(t)$. The input to the system is the electrical signal $E(t)$ sent to the actuator, and the output is the response of the structural system; in this case three accelerations recorded at the three floors $\ddot{x}_1(t)$, $\ddot{x}_2(t)$, $\ddot{x}_3(t)$. All the time-dependent functions are discretized using time steps Δt. We observe that the system in this case is nonlinear. The source of nonlinearity is the combined actuator and structural system. The structural system can either behave linearly or it can behave nonlinearly when the structure suffers damage. There is also a time delay between the actuator receiving the signal and generating the control force.

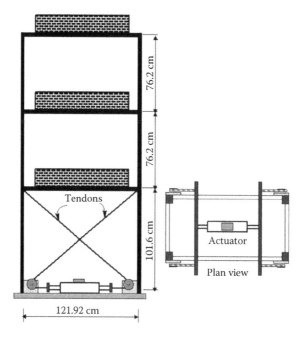

Figure 4.16 The experimental setup for active control of three story frame subjected to earthquake ground acceleration on a shacking table.

We have used neural networks in structural control is several projects (Ghaboussi and Joghataie, 1995; Nikzad et al., 1996; Bani-Hani and Ghaboussi, 1998a; 1998b; Bani-Hani et al., 1999a; 1999b; Joghataie, 1994; Bani-Hani, 1998). In this section, we will describe (Bani-Hani, Ghaboussi, 1998a) with the objective of discussing how neural networks can be used in this type-two inverse problem. We will not go into details and will present only few results. For additional information, the reader is referred to the paper cited above and the doctoral thesis (Ban-Hani 1998).

Although actual experiments were performed on shaking table, we will only describe the part of the study that used the simulated experiment and the structural system was assumed to behave nonlinearly. Only the acceleration in the first floor was assumed to be the system's output. The neurocontroller is shown in Figure 4.17. The output of the neurocontroller is the signal to the actuator. The input of the neurocontroller consists of three parts: the actuator signal at the three previous time steps, the recorded first floor acceleration at the five previous time steps, and the ground acceleration at the two previous time steps. The past information in dynamic systems are needed to characterize the behavior of the system. The number of previous time step data to use as input varies for different problems, and there are no rules on determining the actual number. We observe that all the past data used as the input to the neurocontroller are available and have been recorded.

The training of the neurocontroller requires a model of the system between the input actuator signal and the output first floor acceleration. This model should account for the nonlinearity and the actuator time delay compensation. We chose to train a neural network to learn this type-two inverse problem within the range of interest; this is the emulator neural network shown in Figure 4.18. The input to the emulator neural network contains the actuator signal E, and the output is the predicted first floor acceleration. The input also contains the past history of actuator signal at three time steps and the past history of the first floor acceleration at four time steps.

The emulator neural network is trained by sending random signals to the actuator and recording the response of the system in terms of the first floor acceleration. This training

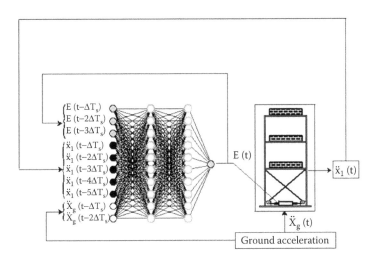

Figure 4.17 Neurocontroller used in control of the motion of the structural system subjected to earthquake ground motion applied on a shaking table. (From Bani-Hani, K. and Ghaboussi, J., *J. Eng. Mech. Division*, 124, 319–327, 1998a.)

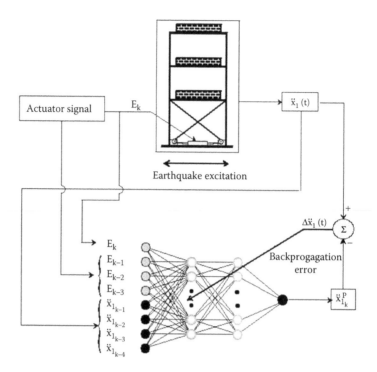

Figure 4.18 Emulator neural network that learns the system identification. (From Bani-Hani, K. and Ghaboussi, J., *J. Eng. Mech. Division*, 124, 319–327, 1998a.)

process is illustrated in Figure 4.18. Two emulator neural networks were trained. Data used in training of the first emulator neural network were generated by assuming that the structural system was linear elastic, whereas data from nonlinear structural system were used in training of the second emulator neural network. The performances of these two emulator neural networks are shown in Figure 4.19.

There are several ways that training data for neurocontroller can be generated. The most obvious way is to directly develop the training data. However, this is not possible in dealing with the dynamical systems, such as structures. In an early application of neurocontrollers, when the objective was to study the role of actuator dynamics and delay compensation, the structure was replaced with a simple mass in the experiment (Nikzad et al. 1996). In that experiment, shown in Figure 4.20, the neurocontroller was trained to simply negate the base motion; if the actuator causes a relative displacement equal to the negative of the base motion, the total displacement of the mass will be zero. However, because of the actuator dynamic and the delay in the actuator response, the total displacement of the top mass will be reduced but not zeroed out. The input to the neurocontroller in this example was the negative of current displacement and past history of displacement of the bottom at seven previous time steps, and the output was the actuator signal. Obviously, in this example a simple control criterion was used and there was no need for an emulator neural network.

In the structural control example, the coupled structure and actuator system is a nonlinear dynamic system and instantaneous control, as in the simple mass example, is not possible. Once the current response is sensed, it is too late to control it. A control criterion is needed that takes into account the forecasted future response of the structure. This is where the emulator neural network comes in, as shown in Figure 4.21.

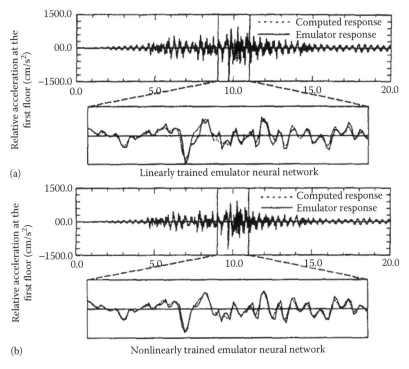

(a)   Linearly trained emulator neural network

(b)   Nonlinearly trained emulator neural network

Figure 4.19 Performance of the trained emulator neural network in two cases: (a) when the structural system is assumed to behave linearly elastic; (b) when the structural system is assumed to be nonlinear. (From Bani-Hani, K. and Ghaboussi, J., *J. Eng. Mech. Division*, 124, 319–327, 1998a.)

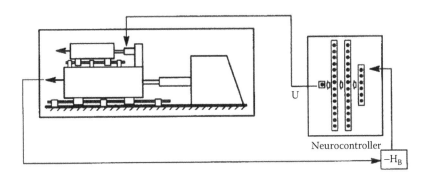

Figure 4.20 Trained neurocontroller sending a signal to the top actuator to control the motion of the top mass relative to motion of the bottom mass. (From Nikzad, K. et al., *J. Eng. Mech. Division*, 122, 966–975, 1996.)

Instead of reducing the response at the next time step, the control criterion is developed to reduce the average structural response over the next few time steps. The emulator neural network is used to determine the response of the structure—in this case the acceleration in first floor—at a few future time steps. Of course, the accuracy of the prediction of the emulator neural network degrades as we move further into future. The predicated response is multiplied by a *prediction validity function*, which has a decreasing value with time steps into future. In earlier studies, the control criterion used data from one emulator neural network (Ghaboussi and Joghataie, 1995; Bani-Hani and Ghaboussi, 1998a; 1998b). In a later

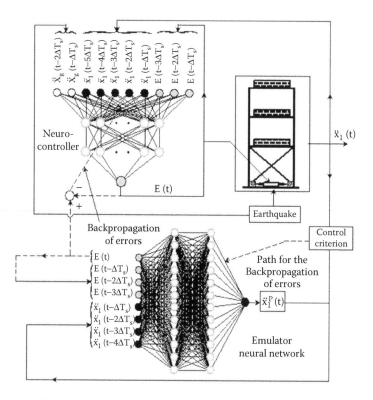

Figure 4.21 Training of the neurocontroller with the trained emulator neural network. (From Bani-Hani, K. and Ghaboussi, J., *J. Eng. Mech. Division*, 124, 319–327, 1998a.)

experimental study of the same three story structural system, three separate emulator neural networks were trained and used in training of neurocontroller; this increased the accuracy of predictions (Bani-Hani et al. 1999a; 1999b).

In training of neurocontroller, the system was subjected to an earthquake ground motion and random signal was sent to the actuator. The sensed response of the structure (acceleration at first floor), along with predicted current and future response of the structure determined from emulator neural network, were sent to the control criterion. Output of the criterion is the error in the current first floor acceleration. This error was backpropagated through the emulator neural network and the neurocontroller. Only the connection weights of the neurocontroller were changed. This process was repeated until the control criterion was satisfied.

We have just presented the overall approach in using neural networks in structural control without going into details. The reader is referred to the references cited for more complete presentation of these studies.

Two neurocontrollers were developed; the first one was trained on linear structure and the second one was trained on nonlinear structure. They were applied in control of the nonlinear structure that could suffer damage subjected to 200% of El Centro earthquake record. The results are presented in Figure 4.22, which shows the damage in the first floor, defined as the reduction in stiffness, for uncontrolled and controlled structures. This clearly demonstrates the effectiveness of the trained neurocontrollers.

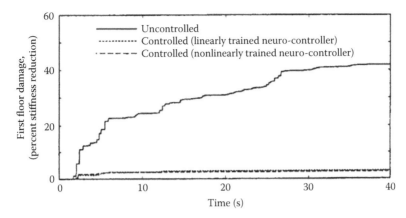

Figure 4.22 Damage in the first floor when the structure is subjected to 200% of El Centro earthquake record. (From Bani-Hani, K. and Ghaboussi, J., *J. Eng. Mech. Division*, 124, 319–327, 1998a.)

4.10 SUMMARY AND DISCUSSION

Theoretically, inverse problems do not have unique solutions. This poses a problem in trying to solve the inverse problems with the mathematically based methods. Inverse problems are solved in nature by learning from the forward problems. Nature has evolved methods of finding the correct locally admissible solution from an infinite number of possible solutions. Nonuniversality, robust imprecision tolerance, and functional nonuniqueness are the fundamental properties of the nature's method of solving the inverse problems and finding the locally admissible solutions. Examples of inverse problems in nature are voice and facial recognition. We solve these problems in a robust and imprecision tolerance manner by learning from the forward problems in a nonuniversal way—we can only recognize a finite number of voices and faces.

Neural networks inherit these capabilities. We have discussed the fact that an inverse problem that mathematically has an infinite number of universal solutions can be learned by a neural network in a locally admissible manner nonuniversally. We have provided the examples of neural networks solving type-one and type-two inverse problems. For type-one inverse problem, we presented a method of using neural networks in generating artificial earthquake accelerograms from the response spectra that are used in engineering practice. In calculating the response spectrum from the earthquake ground acceleration, information is lost. As a consequence, the inverse of determining accelerogram from the response spectrum does not a unique solution. In the method we presented, neural network is trained to learn the inverse problem from data generated from forward problems.

We also presented an application of neural networks in type-two inverse problem of training emulator neural network in structural control. Neural network was trained to learn the nonlinear dynamics of the structure and control system with data generated from forward problems.

The important point in the examples in this chapter is to emphasize the reason why soft computing methods are able to solve these difficult inverse problems that theoretically do not have unique solutions. What enables these capabilities is the fundamental properties of the soft computing methods that was discussed in Chapter 1; imprecision tolerance, nonuniversality, and functional nonuniqueness. These properties lead to methods of robustly finding locally admissible solutions in difficult inverse problems in nature and in engineering.

Chapter 5

Autoprogressive algorithm and self-learning simulation

5.1 NEURAL NETWORK MODELS OF COMPONENTS OF A SYSTEM

In engineering applications, the data for training of neural networks may come from many sources, including experiments, field measurements, recordings during system operations, or computational simulations. In all these types of applications, the data are directly available for training of the neural networks model of the system itself. If we are interested in modeling of the components of the system with neural networks, and the only data available are the stimulus/response (input/output) of the system, then we need to solve an inverse problem. As discussed in the previous chapter, this is a type-two inverse problem; the input and output of the system are known and the model of the system, or model of its components, need to be determined. Autoprogressive Algorithm, also called Self-Learning Simulation, or SelfSim, was developed to solve this class of type-two inverse problems in engineering (Ghaboussi et al., 1998). In most of this class of problems, the data for directly training the neural network model of the components of system are not available; it is either impossible, or difficult to generate those data. However, the stimulus and response of the system itself can be directly measured, and it contains information about the behavior of the components of the system. Autoprogressive Algorithm is able to extract this information in the form of a trained neural network. We will present several examples in this chapter to illustrate the application of the method.

This method was first developed as Autoprogressive Algorithm. In later years, it was also referred to as Self-Learning Simulation, or SelfSim, to emphasize the fact that the neural network model of the components of the system participates in generating the data used in its own training. The underlying method is the same, which ever name is used. In this chapter, we will use the original name of Autoprogressive Algorithm.

The most common example of stimulus/response is applying forces to a structural system and measuring its response in the form of displacements and using these data in the Autoprogressive Algorithm to train the neural network model of the constitutive behavior of material or materials in the structural system.

One example of this class of problems is the modern composite materials. In numerical model of a composite material, the constitutive behavior of the constituent components of the material is needed. They cannot be tested outside the composite material by themselves. For example, in laminated fiber reinforced composites, the constitutive model of the lamina may be needed, but the lamina cannot be tested by themselves. However, the behavior of the specimen of the composite material can be experimentally observed. The constitutive behavior of the lamina can be modeled by neural networks and they can be trained by Autoprogressive Algorithm using the observed and measured response of a structural system of the composite material.

Another example is the in situ measurement of the constitutive properties of geomaterials such as soils and rock in geological formations. Sampling of these materials is often difficult and expensive and often causes the disturbance of the extracted sample. As a result, the measured behavior of the sample in the laboratory may be somewhat different than the in situ behavior of the material. Autoprogressive Algorithm offers the possibility of using the results of an in situ test to determine the constitutive properties of geomaterials without having to extract samples for testing in the laboratory. A good example is the excavations in urban area, where the walls of the excavation are protected and the displacements of the walls and area around the excavation are routinely measured to minimize the ground movements that may affect the adjacent buildings. These data contain information about constitutive behavior of the soils around the excavation that can be extracted by the using Autoprogressive Algorithm.

There are many similar examples of engineering problems where some properties of the constituent materials or other components are needed, but it is difficult or impossible to measure them directly outside the system. Another area of the application Autoprogressive Algorithm is modeling the constitutive properties of soft tissue in the field of biomedicine. We will discuss some examples applications of Autoprogressive Algorithm in biomedicine.

Autoprogressive Algorithm was developed to enable the training of the neural network models of components of system by observation and recording of the stimulus/response (input/output) of the system itself. Autoprogressive Algorithm requires that the neural network models of the components of the system be used in the numerical simulation of the system. Autoprogressive Algorithm is an iterative process of retraining of the neural network models of the components of the system, to enforce the condition that the stimulus/response of the numerical model be made to match the observed stimulus/response of the system within a predefined error tolerance

5.2 AUTOPROGRESSIVE ALGORITHM

The word autoprogressive is used to signify that the neural network model of the components of a system is an integral part of the process of generating the progressively improved data for training of the neural network. Neural networks may be initially pretrained with some simple idealized models of the behavior of the components, such as linearly elastic behavior in the case of constitutive models. In Autoprogressive Algorithm, computational simulations of the stimulus/response of the system are performed with the behavior of its components represented by the neural networks. Initially, the neural networks do not represent the component behavior well, and the results of the computational simulation of the system response will differ from the recorded system response. Neural networks are trained with new data produced by the computational simulation and the process is repeated. This iterative process would yield increasingly improved training data for the neural networks, which learn to represent the component behavior with increasing accuracy. When the neural networks are fully trained, the response of the system from the computational simulation will match the recorded response to within an error tolerance.

A typical system, shown schematically in Figure 5.1, has many components or agents. Also shown in this figure is the computational model of the system, where the behavior of its components is modeled with neural networks symbolically expressed in the following equation:

$$AO = NN[AI; \quad] \tag{5.1}$$

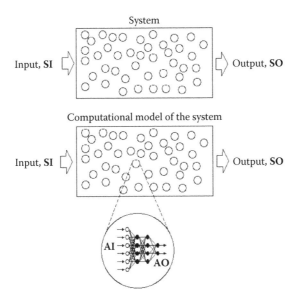

Figure 5.1 A complex system whose components are modeled by neural networks.

The input vector for the neural network is **AI**, and the output vector is **AO**. It is also important to note that the stimulus–response behavior of the components of a system may not be deterministic and the neural network models may need to incorporate some features of stochastic behavior in their responses.

We further assume that the vector-valued function f represents the computational model of the system, which relates the output (response) of the system, **SO**, to its input (stimulus), **SI**

In general, f is nonlinear and may be partly a stochastic function.

$$\mathbf{SO} = \mathbf{f}\left(\mathbf{SI},\ \mathbf{NN}[\mathbf{AI};\quad]\right) \tag{5.2}$$

The simplest case would arise from the assumption that all the components or agents of the system are identical and that a single neural network can therefore represent their behavior. In the earlier versions of the Autoprogressive Algorithm applied in computational mechanics, this simplified model with one type of component was used. However, this assumption will not be valid in most systems where there may be many different types of components and each would be modeled with a different neural network. This is not a restriction in Autoprogressive Algorithm and in many applications more than one neural network has been used to model the components of the system.

In recent application of Autoprogressive Algorithm in the biomedical imaging of the soft tissue material properties, *Cartesian neural network material models* are used. This type of neural network not only learns to model the behavior of the soft tissue, it also learns the spatial variation the tissue material properties over a region of interest. It is called Cartesian because spatial coordinates are part of the input to neural network. This approach addresses the problem that the spatial variation of tissue properties, and a number of neural networks needed, are not known a-priori.

Autoprogressive Algorithm requires that the neural network models of the components of the complex system be initially pretrained to idealized and simple approximation of their actual behavior. The pretrained neural networks used in the computational model of the complex system define the initial behavior of the system.

Autoprogressive Algorithm involves computational simulation of the system's response to input stimuli. Since almost all systems behave nonlinearly, the simulation is performed incrementally. In each increment, several iterations are performed to continue training of the neural network model of the components of the system. In each iteration, two simulations are performed. In the first simulation, the system is subjected to the input stimuli and in the second simulation the system is subjected to the measured output response. We observe that the second simulation is a type-one inverse problem, discussed in the previous chapter. In computational mechanics, both simulations are possible and we present examples to illustrate this point. However, in some systems it may be difficult or impossible to directly perform the inverse second simulation.

To continue discussion of the Autoprogressive Algorithm, we will assume that both simulations are possible. In the nth iteration, the following two simulations are performed. In both simulations, the component properties are represented by the latest version of the trained neural network, or neural networks.

Simulation 1: The system is subjected to the measured input stimuli $\mathbf{SI}_n$ and the response of the system $\overline{\mathbf{SO}}_n$ is computed that differs from the measured response $\mathbf{SO}_n$.

Simulation 2: The system is subjected to the measured output response $\mathbf{SO}_n$ and the input of the system $\overline{\mathbf{SI}}_n$ is computed that differs from the measured input $\mathbf{SI}_n$.

The primary working mechanism behind the Autoprogressive Algorithm is that these two simulations produce increasingly more accurate data for training of the neural network. In any system, there are relationships between the input to the system and the input to its components as well as relationship between the output of system and the output of its components. These relationships are governed by fundamental laws. Simulation 1 is closer to satisfying the law that governs the relations between $\mathbf{SI}_n$ and $\mathbf{AI}_n$. Therefore, it is reasonable to assume that the computed input to its components $\overline{\mathbf{AI}}_n$ is closer to the actual input to the neural network. Similarly, simulation 2 is closer to satisfying the law that governs the relations between $\mathbf{SO}_n$ and $\mathbf{AO}_n$. Therefore, it is reasonable to assume that the computed output of its components $\overline{\mathbf{AO}}_n$ is closer to the actual output to the neural network. $\overline{\mathbf{AI}}$ and $\overline{\mathbf{AO}}$ are added to the training data and the neural network is retrained with the new data.

The process of two simulations and retraining of the neural network is iteratively continued within the nth increment until the output of the simulation 1 and input of the simulation 2 approach the measured input and output to within a prescribed error tolerance

$$\begin{cases} \delta\mathbf{SO}_n = \overline{\mathbf{SO}}_n - \mathbf{SO}_n \\ \delta\mathbf{SI}_n = \overline{\mathbf{SI}}_n - \mathbf{SI}_n \end{cases} \tag{5.3}$$

$$\begin{cases} \dfrac{\left\|\delta\mathbf{SO}_n\right\|}{\left\|\mathbf{SO}_n\right\|} \leq \varepsilon \\ \\ \dfrac{\left\|\delta\mathbf{SI}_n\right\|}{\left\|\mathbf{SI}_n\right\|} \leq \varepsilon \end{cases} \tag{5.4}$$

This process is continued for all the increments. This is called a *pass*. Several passes may be required until neural networks learn the actual behavior of the components of the system and simulations 1 and 2 produce the same results that are close to the measured data to within an error tolerance.

5.3 AUTOPROGRESSIVE ALGORITHM IN COMPUTATIONAL MECHANICS

5.3.1 Neural network constitutive models of material behavior

Modeling of the nonlinear constitutive behavior of materials is an important part of the numerical simulation of structural systems with finite-element (FE) or finite-difference methods. Neural networks can be used to model the nonlinear constitutive behavior of materials as has been shown in Chapter 3. Plasticity-based constitutive models arrive at an incremental relation between strains increments and stress increments, as in the following equation:

$$\Delta\sigma = \mathbf{C}^{ep}\Delta\varepsilon \tag{5.5}$$

In this equation, $\mathbf{C}^{ep}$ is the elastoplastic constitutive matrix.

Neural networks can also be trained to represent the nonlinear constitutive behavior of materials and a typical neural network material model would have the following form:

$$\Delta\sigma_n = \mathbf{NN}\left[\Delta\varepsilon_n, \sigma_n, \varepsilon_n, \ldots, \sigma_{n-k}, \varepsilon_{n-k} : \quad\right] \tag{5.6}$$

The input to this neural network also includes the history points that consist of the current state of stresses and strains and the states of stresses and strains at k previous steps. The history points are used to account for the path dependence of the constitutive behavior of materials. In Chapter 3, we showed how k can be determined by using the concept of nested adaptive neural networks. In Chapter 3 we also discussed an alternative form of accounting for path dependence and hysteretic behavior of material by using strain energy, as in the following equation:

$$\sigma_n = \mathbf{NN}\left(\varepsilon_n, \sigma_{n-1}, \varepsilon_{n-1}, \xi_{n-1}, \Delta\eta_\varepsilon ; \quad\right)$$

$$\begin{cases} \xi_{n-1} = \sigma_{n-1}\,\varepsilon_{n-1} \\ \Delta\eta_\varepsilon = \sigma_{n-1}\left(\varepsilon_n - \varepsilon_{n-1}\right) \end{cases} \tag{5.7}$$

Later in this chapter, we will present and discuss examples of using this type neural network material model in Autoprogressive Algorithm.

In the early applications of neural networks in constitutive modeling discussed in Chapter 3, the neural network material models were trained directly with the data from material tests. Constitutive modeling from the material tests is a type-two inverse problem. In this case, the input (applied stresses) and the output (measured strains) are known and the system, which in this case is the constitutive model, needs to be determined. In the mathematically based methods, this is accomplished by developing a mathematical model that closely matches the observed behavior. Of course, the mathematical models have to conform to the conservation laws of mechanics, described in detail in the book by Ghaboussi et al. (2017). As we have discussed before, in using neural networks, we concentrate on the information on the material behavior contained in the data generated from the material tests. The learning capabilities of neural networks provide the means for extracting and storing the information on material behavior directly from data generated by the material experiments.

The advantage of extracting the information on the material behavior directly from the experimental results is that there is no need for idealization. Moreover, as the data come from the observed behavior of the material itself, it is safe to assume that it conforms to the conservation laws of mechanics. There are also disadvantages in training neural network material models directly from material tests. Material tests are designed to represent a material point. Consequently, the state of stresses and strains within the sample must be as uniform as possible. The sample is subjected to a stress path and all the points within the sample are assumed to follow the same stress path. Therefore, the data generated from a material test have information about the material behavior only along that stress path. The data from one material test are not sufficient to train a robust neural network material model that has the knowledge of the material behavior in the stress and strain space within the range of interest. Information on the material behavior over the whole region of interest in stress space is needed to train a robust neural network constitutive model with generalization capabilities. This requires a series of specially designed material tests with stress paths that reasonably cover the region of interest in the stress space. This is not practical in most cases.

Material tests are not the only source of information on the behavior. There are many other potential sources of data that contain information on material behavior. The measured response of a system subjected to a known excitation contains information on the constitutive behavior of the material (or materials) in that system. An example is a structural test; the measured displacements of a structural system that is subjected to known forces contain information on the constitutive properties of the materials within the structure. This is a more complex inverse problem than the material tests; the forces and displacement are the known input and output of the system and the constitutive properties of the materials within the structural system need to be determined. Unlike material tests, which ideally induce a uniform state of stress within the sample, structural tests induce nonuniform states of stresses and strains within the sample. Since points in the specimen follow different stress paths, a single structural test potentially has far more information on the material behavior than a material test in which all the points within the sample follow the same stress path.

Extracting the information on the material properties from structural tests is an extremely difficult problem with the conventional mathematically based methods. This is probably the reason why it has not been attempted successfully in the past. On the other hand, soft computing methods are ideally suited for this type of difficult inverse problem. The learning capabilities of a neural network offer a solution to this problem. Autoprogressive Algorithm is a method for training a neural network to learn the constitutive properties of materials from structural tests, and it was first introduced in Ghaboussi et al. (1998).

5.3.2 Training of neural network material models from structural tests

The Autoprogressive Algorithm requires the measurement of applied forces and resulting displacements in the structural test. However, training of the neural network material model requires the determination of the output error of the neural network for any input pattern, that is, the error in stress for a given strain input. This detailed local information is not directly available from the global structural responses recorded during the structural test. To generate the detailed local stress–strain information that is needed for training purposes, the neural network material model that is to be trained is itself used in a nonlinear FE simulation of the structural test.

Prior to autoprogressive training, the neural network material model needs to be initialized by pretraining it on a dataset $\mathbf{A}_e$ representing an idealized constitutive model such as a linear elastic model.

$$\mathbf{A}_e = \left[\varepsilon_l^e, \sigma_l^e \; ; l = 1, \cdots, \mathrm{L} \right] \tag{5.8}$$

This dataset is also incorporated into the training data generated during the incremental FE simulation of the structural test. Iterations are performed in each load increment, and each iteration consists of a pair of FE analyses. This dual FE analysis is at the heart of the Autoprogressive Algorithm.

In the first FE analysis, a measured force increment is applied and a routine forward analysis is done, using the current neural network material model in the analysis. Since the neural network material model is not completely trained at this stage, the displacement increment produced by the first analysis will not match the measured displacement increments. The magnitude of the displacement error (the difference between the measured and the computed displacements) is related primarily to how well the neural network material model simulates the actual material behavior.

In the second, FE analysis, the measured structural displacements are applied as imposed displacement boundary condition. There are two ways of using the results of the second FE analysis. We will describe both approaches.

The first approach was used in the original development of the Autoprogressive Algorithm reported in Ghaboussi et al. (1998). In this approach, the difference between the first and second FE analyses is treated as representing the error field needed for training of the neural network material model. The difference between stresses from the first and second analyses is used as the output errors of the neural network material model. They are added to the training data and used in retraining the neural network material model.

The second approach was used in the later applications of Autoprogressive Algorithm. This approach is based on the following observations: in any FE analysis equilibrium and compatibility are satisfied. Equilibrium establishes the relation between the forces at the structural degrees of freedom and stresses within the elements. Compatibility establishes the relation between displacements at the structural degrees of freedom and strains within the elements. We can make the following observation on the two FE analyses.

FE analysis one (FEA1): Since the actual forces are applied, it is reasonable to assume that in FEA1 the equilibrium will be closer to the actual equilibrium in the structural system. Therefore, stresses are closer to the actual stresses.

FE analysis two (FEA2): Since the actual measured displacements are applied, it is reasonable to assume that in FEA2 the comparability will be closer to the actual compatibility in the structural system. Therefore, strains are closer to the actual strains.

Stresses from FEA1 and strains from FEA2 are added to the training data and the neural network material model is trained with the new dataset. The trained neural network is then used in the next iteration or the next load step.

Both approaches described above work and give similar results. We will continue discussion of the Autoprogressive Algorithm concentrating on the second approach since it was used in later and more current applications while recognizing that the first approach is equally valid and some reader may prefer to use that.

In each load increment, the dual FE analyses are carried out iteratively, until the neural network material model has satisfactorily learned all the training cases generated up to and including the current load increment. Application of all the load increments, up to the maximum applied forces, constitutes one *pass*.

Because the neural network is at first only partially trained, the training cases that are derived from the structural analysis are not initially consistent with the actual stress–strain response of the material. For this reason, the complete training of the neural network material model may require several passes. Eventually, as the neural network material model gradually evolves with additional training and more closely represents the actual material behavior, both FE analyses produce the results that closely match the measured forces and displacements.

The training algorithm is described in the following steps, which describe the iterative operations that are carried out at the nth load increment. It is assumed that at the beginning of the load step, the neural network material model has been partially trained and is capable of approximately representing the nonlinear material behavior up to that point. However, it is not yet capable of accurately matching the real material behavior.

At the nth load step, the load increment ΔP_n is applied and the following FE equations are solved iteratively.

$$P_n = P_{n-1} + \Delta P_n \tag{5.9}$$

$$K_t \Delta U_n = P_n - I_{n-1} \tag{5.10}$$

In Equation 5.10, the internal resisting force vector, I_{n-1}, is computed from the stresses at the end of the previous increment.

$$I_{n-1} = \sum \int B^T \left\{ \sigma_{n-2} + \Delta\sigma_{n-1} \, NN\left[\Delta\varepsilon_{n-1}, \sigma_{n-2}, \varepsilon_{n-2}, \cdots : \quad \right] \right\} dv \tag{5.11}$$

The behavior of the structure under this load increment is in general nonlinear, and Equation 5.10 is solved by an iterative method such as modified Newton–Raphson method (Ghaboussi and Wu, 2016). The primary difference with a standard nonlinear FE analysis is that a two-step analysis is required in each iteration.

5.3.2.1 FEAI

This is a standard FE analysis to determine the displacement increments resulting from the applied load increment. The following equations describe the first step forward analysis in the jth iteration in the nth load increment.

$$K_t^{j-k}\delta U_n^i = P_n - I_{n-1} - \delta I_{n-1}^{j-1} \tag{5.12}$$

$$K_t^{j-k} = \sum \int B^T D_t^{j-k} B \, dv \tag{5.13}$$

The constitutive behavior of the material at the integration points in the FEs are represented by the stress–strain matrix developed from the neural network material model, as described in Chapter 3, Section 3.9.

$$\delta I_{n-1}^{j-1} = \sum \int B^T \delta\sigma_n^{j-1} NN\left[\delta\varepsilon_n^{j-1}, \sigma_{n-1}, \varepsilon_{n-1}, \cdots \quad : \quad \right] dv \tag{5.14}$$

It can be seen from Equation 5.14 that the stresses and strains used as input to the neural network material model are the values at the end of the previous load increment. As mentioned earlier, the tangent stiffness matrix is not needed if the conjugate gradient method is used. Otherwise, in direct solution method, the tangent stiffness matrix is usually updated infrequently. Equation 5.13 indicates that the tangent stiffness matrix was last updated at the iteration number $j-k > 0$.

The neural network constitutive models in Equation 5.14 were last updated and retrained at the end of the previous iteration (if this is the first iteration, they were last retrained at the end of the previous increment). The training dataset for these neural networks include the cases generated at the end of the previous iteration $\mathbf{A}_n^{j-1}$ described by the following equation:

$$\mathbf{A}_n^{j-1} = \left[\epsilon_{nm}^{j-1}, \sigma_{nm}^{j-1} \; ; \; m = 1, \cdots, M \right] \tag{5.15}$$

In this equation, M is total number of the Gaussian integration points in all the elements.

In addition to the training data generated at the end of the previous iteration, the complete training dataset at this point $\mathbf{TD}_n^{j-1}$ also includes a moving window of the training data generated at the end of the previous k load steps (the reasons will be explained later), and the pretraining dataset $\mathbf{A}_e$.

$$\mathbf{TD}_n^{j-1} = \left[\mathbf{A}_e, \mathbf{A}_{n-k}, \cdots, \mathbf{A}_{n-1}, \mathbf{A}_n^{j-1} \right] \tag{5.16}$$

The structural equilibrium equation in Equation 5.12 is solved to determine the iterative incremental displacements $\delta \mathbf{U}_n^j$. The corresponding strain increments are then computed at element integration points, and the stress increments are determined by a forward pass through the neural network material model.

$$\delta \epsilon_n^j = \mathbf{B} \delta \mathbf{U}_n^j \tag{5.17}$$

$$\delta \sigma_n^j = \delta \sigma_n^j \mathbf{NN} \left[\delta \epsilon_n^j, \sigma_{n-1}, \epsilon_{n-1}, \cdots \quad : \quad \right] \tag{5.18}$$

The incremental values from the dual analysis in the jth iteration are used to update the incremental displacements, strains and stresses for the nth load increment.

$$\begin{cases} \Delta \mathbf{U}_n^j = \Delta \mathbf{U}_n^{j-1} + \delta \mathbf{U}_n^j \\ \Delta \epsilon_n^j = \Delta \epsilon_n^{j-1} + \delta \epsilon_n^j \\ \Delta \sigma_n^j = \Delta \sigma_n^{j-1} + \delta \sigma_n^j \end{cases} \tag{5.19}$$

Note that the total displacements, strains, and stresses are not updated, until the iterative process, including the retraining of the neural network material model, has converged.

5.3.2.2 FEA2

Since the neural network is not fully trained at this stage, the displacements from FEA1 may not match the measured displacements. In FEA2, the measured incremental displacements $\delta \bar{\mathbf{U}}_n^j$ are applied at the degrees of freedom where the displacements are measured. The second FE analysis provides the displacements at all the other degrees of freedom of the structure $\delta \mathbf{U}_n^j$. The following equation describes the second FE analysis where the structural stiffness equations have been partitioned to separate the degrees of freedom with measured displacements from all the other degrees of freedom.

$$\begin{bmatrix} \mathbf{K}_t^{j-k}{}_{aa} & \mathbf{K}_t^{j-k}{}_{ab} \\ \mathbf{K}_t^{j-k}{}_{ba} & \mathbf{K}_t^{j-k}{}_{bb} \end{bmatrix} \begin{Bmatrix} \delta \mathbf{U}_n^j \\ \delta \bar{\mathbf{U}}_n^j \end{Bmatrix} = \begin{Bmatrix} \mathbf{P}_{n-1} \\ \bar{\mathbf{P}}_{n-1} + \delta \bar{\mathbf{P}}_n^j \end{Bmatrix} - \begin{Bmatrix} \mathbf{I}_{n-1} + \delta \mathbf{I}_{n-1}^{j-1} \\ \bar{\mathbf{I}}_{n-1} + \delta \bar{\mathbf{I}}_{n-1}^{j-1} \end{Bmatrix} \tag{5.20}$$

The partitioning of the stiffness matrix in Equation 5.20 is symbolic, and it is not necessary in practice. Note that the unknowns in Equation 5.20 are the displacement δU_n^j at the degrees of freedom with unmeasured displacements, and $\delta \bar{P}_n^j$, the forces at the degrees of freedom with measured displacements.

The strain increments at the integration points are computed from the displacement of the second FE analysis.

$$\delta \varepsilon_n^j = \mathbf{B} \begin{Bmatrix} \delta U_n^j \\ \delta \bar{U}_n^j \end{Bmatrix} \tag{5.21}$$

The incremental values from the second FE analysis in the jth iteration are used to update the incremental displacements and strains for the nth load increment.

$$\begin{cases} \Delta U_n^j = \Delta U_n^{j-1} + \delta U_n^j \\ \Delta \varepsilon_n^j = \Delta \varepsilon_n^{j-1} + \delta \varepsilon_n^j \end{cases} \tag{5.22}$$

5.3.2.3 Retraining phase of the autoprogressive algorithm

After completing the dual FE analyses FEA1 and FEA2, the training dataset for the neural networks is updated, and the neural network constitutive models are retrained. In the new dataset, the strains are updated with the increments from the second FE analysis, whereas the stresses are updated with the increments from the first FE analysis.

$$\mathbf{A}_n^j = \left[\varepsilon_{nm}^{j-1} + \delta \varepsilon_{nm}^j, \sigma_{nm}^{j-1} + \delta \sigma_{nm}^j ; m = 1, \ldots, M \right] \tag{5.23}$$

At each new iteration, the training cases, $\mathbf{A}_n^j$, replace in the training database those that were generated in previous local iterations at load step n, $\mathbf{A}_n^{j-1}$. Iterations are continued until the convergence criterion, described later, is satisfied; that is, until FEA1 and FEA2 produce the same results, as close to the measured force and displacement, within a specified error tolerance. Once this satisfactory agreement is achieved at load step n, the process is repeated for load step (n + 1), and so on, until the full range of applied loads is covered. The term *load pass* is used to denote the application of this algorithm over the full range of applied loads. As the Autoprogressive Algorithm proceeds from one load increment to the next, a set of stress–strain training pairs is collected at each load step from the converged solution. At the conclusion of one load pass, the entire training database consists of the pretraining cases, $\mathbf{A}_e$ plus one set of training cases for each load step. Because the connection weights are updated periodically during the load pass, these different sets of training cases actually correspond, in effect, to different material models. In general, the most recently collected sets more closely represent the actual material behavior. Therefore, some strategy must be adopted to either retire or correct the older training sets. One effective strategy that has been developed and used effectively is to use a moving window to select only those training sets that were collected over a specified small number (k) of prior load steps.

$$\mathbf{TD}_n^j = \left[\mathbf{A}_e, \mathbf{A}_{n-k}, \cdots, \mathbf{A}_{n-1}, \mathbf{A}_n^j \right] \tag{5.24}$$

This dataset is used to determine the input and output to the neural networks. History points at previous load steps and current state of stresses and strain at the beginning of the load step are also available in this dataset. In more complex material modeling situations that involve multiaxial path-dependent behavior, where it is necessary to include stress path information in the neural network input, additional measures may be necessary to ensure that the stress path information is consistent with the current material model (Zhang, 1996).

5.3.2.4 Convergence of iterations

The iterations are continued until satisfactory convergence according to some predefined convergence criterion is satisfied. As the neural network constitutive models learn to more closely approximate the material behavior, the displacement increments get smaller. The obvious convergence criterion would therefore be in terms of the norm of the difference between displacement vectors at two consecutive iterations:

$$\frac{\left\| \delta U_n^j - \delta U_n^{j-1} \right\|}{\left\| \Delta U_n^j \right\|} \leq \varepsilon \qquad (5.25)$$

When the convergence criterion is satisfied, then the iterations for the nth load step are terminated and the total displacements, strains and stresses are updated.

$$\begin{cases} U_n = U_{n-1} + \Delta U_n^j \\ \varepsilon_n = \varepsilon_{n-1} + \Delta \varepsilon_n^j \\ \sigma_n = \sigma_{n-1} + \Delta \sigma_n^j \end{cases} \qquad (5.26)$$

5.3.2.5 Multiple load passes

The process described earlier for one load step is repeated until the full range of the load steps has been applied. This is referred to as a *pass*. It is important to emphasize that the stress–strain training cases that are derived from this algorithm are not initially representative of the actual stress–strain response of the material. For this reason, the complete training of the neural network material model may require several load passes. With additional load passes, the difference between the results form FEA1 and FEA2 become smaller and eventually, convergence is declared.

In the next section, we will present an illustrative example of a hypothetical simple structure to demonstrate the steps in the autoprogressive training of neural network constitutive models.

5.4 ILLUSTRATIVE EXAMPLE

A plane stress cantilever structure is chosen to demonstrate the Autoprogressive Algorithm for training of the neural network constitutive models (Hashash et al., 2009). The FE model is shown in Figure 5.2. A horizontal force is applied at the top left-hand side. The force is gradually increased to activate the nonlinear elastoplastic response of the structural system.

In the simulated experiment, the material properties are modeled with Von-Mises plasticity with hardening. The response of the simulated experiment is shown in Figure 5.3 in terms of applied force versus the horizontal displacement of the top middle point.

A simple neural network was used to represent the material behavior. Neural network model of elastoplastic material behavior requires that the path dependence be accounted for in the neural network architecture. In Chapter 3, we have seen that one method of

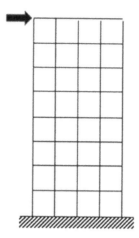

Figure 5.2 Example structure plane stress cantilever.

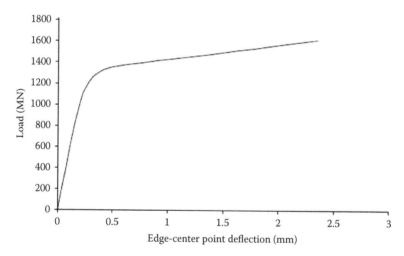

Figure 5.3 Force–displacement response of the simulated experiment.

accomplishing this is by using history points as input. In this example, our purpose is to illustrate how Autoprogressive Algorithm works. For this purpose, we have chosen a simple neural network shown in the following equation, without the history points, mainly because there is very little unloading and reloading in this example.

$$\sigma_n = \sigma_n \, \text{NN}[\varepsilon_n; 3|10\,|10|3] \tag{5.27}$$

A preparatory step is needed prior to starting the Autoprogressive Algorithm. Rather than assigning random initial connection weights to the neural network constitutive model, it is required to initialize the neural network material model by pretraining it on a dataset generated from a linear elastic constitutive model. The linearly elastic dataset is retained in the training data for the Autoprogressive Algorithm generated during the incremental FE simulation of the structural test. In this example, pretraining datasets were generated randomly, with strains in the range (−0.001, 0.001). This constitutes the pretraining database.

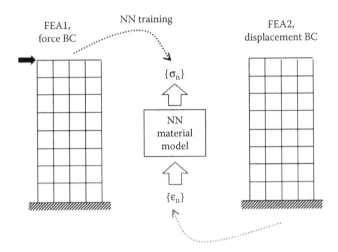

Figure 5.4 Dual finite-element analyses and retraining of neural network material model in Autoprogressive Algorithm.

In contrast to the Autoprogressive phase of training, this pretraining procedure is entirely conventional, in that, all the stress–strain training cases represent true material behavior (in this case, linear elastic behavior).

In the autoprogressive training, the applied force was used in FEA1. In the first 10 passes, only the vertical and horizontal displacements at the top middle point were used the FEA2. In passes 11 and 12, displacements at all the nodes along the outer boundary were used in FEA2. The FEA1 and FEA2 and training of the neural network material model are schematically demonstrated in Figure 5.4. During the Autoprogressive Algorithm, the total force was applied in 10 increments and a number of iterations were performed in each load increment. In each iteration, the dual FE analyses FEA1 and FEA2 were performed with the latest version of the trained neural network representing the constitutive material behavior. Stresses from FEA1 and strains from FEA2 were added to the training dataset that was used in retaining of the neural network material model, shown in Figure 5.4. This iterative process is continued until satisfactory convergence is attained in the current load step. The whole iterative process is repeated for all the 10 load steps, completing a pass. As many passes are performed as is deemed sufficient for the satisfactory training of the neural network constitutive model, in this case 12 passes.

Before autoprogressive training, the pretrained neural network material model that was trained with linearly elastic data was used in a forward FE analysis. Also, at the end of each pass, the trained neural network material model was used in a forward FE analysis of the structural system. The stress–strain relations from these forward analyses are compared with the stress–strain relations at the same points from the simulated experiment with elastoplastic material model. These comparisons illustrate the process of gradual learning of the material behavior during the autoprogressive training. Stress–strain relations at two points from forward analyses after pretraining and at five passes are shown in Figures 5.5 and 5.6.

The stress–strain relations with the pretrained neural network material model are linearly elastic, as expected. It is clear that the neural network material model gradually learns the nonlinear material properties. After 12 passes, it closely matches the stress–strain relation from the simulated experiment with elastoplastic material properties.

We can see the similar pattern in the response of the structural system in forward analyses with the trained neural network compared with the response of the simulated

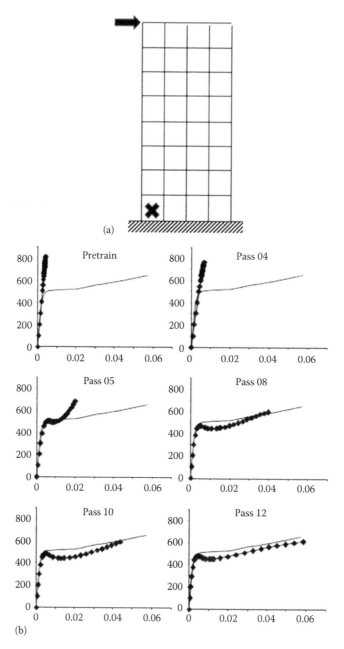

Figure 5.5 Location of the horizontal stress–strain relations: (a) horizontal stress–strain relations from forward analyses with the pretrained NN and (b) trained NN at the end of five passes compared with the stress–strain relation from simulated experiment with the elastoplastic material model. (From Hashash, Y.M.A. et al., *Inverse Probl. Sci. Eng.*, 17, 35–50, 2009.)

experiment. Figure 5.7 shows the relation between the applied force and the horizontal displacement of the top middle point.

It should also be noted that, strictly speaking, the type of comparison shown in the above figures addresses only how consistent the trained neural network material model is with the simulated experimental data on which it was trained. Neural network learns about the

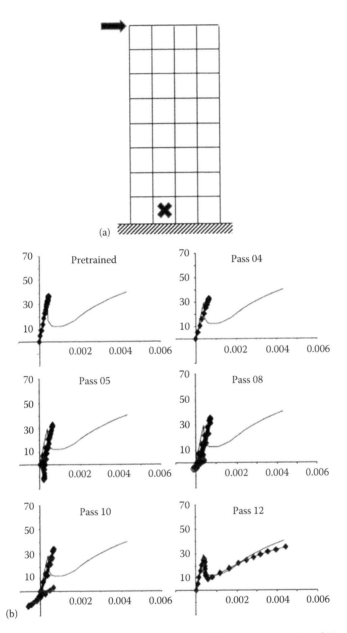

Figure 5.6 Location of the shear stress–strain relations: (a) Shear stress–strain relations from forward analyses with the pretrained NN and (b) trained NN at the end of five passes compared with the stress–strain relation from simulated experiment with the elastoplastic material model. (From Hashash, Y.M.A. et al., *Inverse Probl. Sci. Eng.*, 17, 35–50, 2009.)

material behavior only from the experiment on structural system; how well it has learned the material behavior depends on how information rich is the experiment about the material behavior. This example is only for the purpose of illustration and it is likely that neural network has not learned the complete material behavior to generalize to predict the response of other structural systems with the same material where there may be unloading

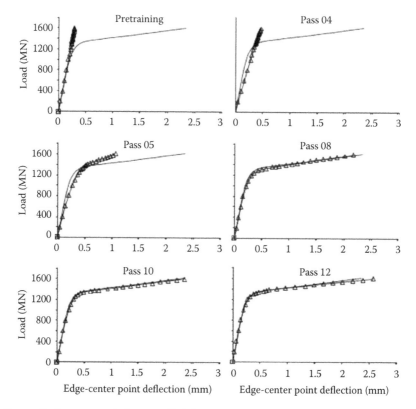

Figure 5.7 Force–displacement relation—applied force versus horizontal displacement at the top middle node—during the autoprogressive training of the neural networks material model compared with the simulated experiment. (From Hashash, Y.M.A. et al., *Inverse Probl. Sci. Eng.*, 17, 35–50, 2009.)

and reloading. Some structural systems are more information rich about the material behavior than others. The training of the neural network material models can be improved by simultaneously applying the Autoprogressive Algorithm to multiple structural systems.

A reasonable evaluation of the trained neural network material model would require using it in other structural systems, constructed of the same material but of different geometrical form, which were not included in the training dataset.

5.5 AUTOPROGRESSIVE ALGORITHM APPLIED TO COMPOSITE MATERIALS

5.5.1 Laminated composite materials

This was the first major application of the Autoprogressive Algorithm that was used in modeling of nonlinear behavior and damage in composite materials. This example is based on the doctoral dissertation of Dr. Mingfu Zhang (Zhang, 1996; Ghaboussi et al., 1998).

Fiber-reinforced laminated composite materials are made up of many layers, and each layer consists of fibers in a certain direction and a matrix material between the fibers (Aboudi, 1991). Under increasing load, laminated composites may develop several different modes of damage such as transverse cracking, shearing or crushing of the matrix, breaking or microbuckling of fibers followed by kink band formation, fiber-matrix debonding, and interply delamination.

Such complex behavior is difficult to characterize via conventional mathematical modeling approaches. Neural networks appear to offer an attractive alternative, particularly for the purposes of nonlinear analysis of composite structural systems. In the neural network approach that is described in this section, the view is taken that damage that occurs at the micromechanical level is implicitly reflected in the effective stress–strain response; thus, specific modes of damage are neither explicitly modeled nor identified. A corollary to this is, of course, that a fully trained neural network material model must be trained on a spectrum of structural test data in which all the important damage modes are sufficiently manifested.

5.5.2 Test setup and specimen

In the example discussed here, experimental results on laminated graphite/epoxy structural plates containing an open hole, reported by Lessard and Chang (1991), are used to train a neural network material model for the composite material. A series of test results for plates with several different layups, including cross-ply, angle-ply, and quasi-isotropic ply orientations, made from T300/976 graphite/epoxy (unidirectional) pre-preg tapes, are reported. Two angle-ply layups are selected for purposes of this example: (1) a $[(\pm 45^\circ)_6]_s$ plate (24 individual plys, with fiber directions rotated alternately with respect to the load direction, and symmetrically placed with respect to the mid-surface of the plate); and (2) a $[(\pm 30^\circ)_6]_s$ plate. The plates were loaded in compression as indicated schematically in Figure 5.8. Experimental results for the $[(\pm 45)_6]_s$ plate are used for training, and the trained neural network material model is then used in a forward analysis, that is, predictive analysis of the $[(\pm 30^\circ)_6]_s$ plate.

5.5.3 Finite-element model of the specimen

The FE model of one-fourth of the specimen is shown in Figure 5.8. Only one-fourth of the test specimen needed to be modeled due to double symmetry. The in-plane compression load in the test was applied through a rigid loading head. This is modeled by applying a

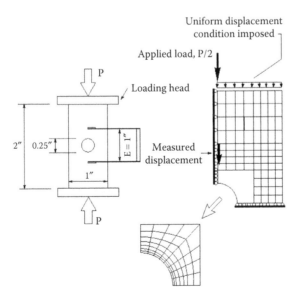

Figure 5.8 The specimen and the test setup for laminated composite plate with a central hole. (From Lessard, L. and Chang, F-K., *J. Compos. Mater.*, 25, 44–64, 1991.)

constraint in the FE model. The measured structural response is the relative displacement across a 1-in gage length, indicated by the dimension labeled "E" in Figure 5.8.

In formulating the neural network approach to this problem, as in any conventional approach to the analysis of composite structures, it must first be decided at what level of microstructural detail the material behavior is to be described: that is, what is the fundamental "material"? Modeling approaches that can be described as micromechanical, and that are viable for structural analysis, have been proposed (Pecknold and Rahman, 1994). In these approaches, the fiber and matrix properties are explicitly recognized. Here, however, a macromechanical approach is adopted for modeling the behavior of the composite material; the fundamental material building block is considered to be an individual ply, which is approximately 0.005 in thick and which consists of unidirectional graphite fibers embedded in an epoxy resin. The stress and strain variables with which the neural network material model operates are therefore chosen to be the effective stresses and strains acting on a single unidirectional ply. Thus, once the neural network material model for a unidirectional ply is successfully trained, it can be used in forward analysis of other composite structures made from the same type of pre-preg tape.

The structural properties of a composite laminate can be synthesized from the equivalent material properties of a single unidirectional ply using either Classical Lamination Theory (CLT) for plate or shell structural models, or 3D lamination theory for three-dimensional FE structural models. A 2D representation of the unidirectional ply properties was chosen; thus, CLT was used in synthesizing the properties for structural analysis. The different levels of hierarchy in the structural model of the composite plate are illustrated in Figure 5.9.

There is some experimental evidence that this particular type of composite material is not strongly path dependent. Therefore, the assumption is made that the material behavior can be approximated as nonlinear elastic. This assumption permits the choice of a very simple neural network representation; three current 2D strain components are input and the three corresponding stresses are output. Two hidden layers, with adaptive architecture determination, are used. The adequacy of this simple representation is discussed further later.

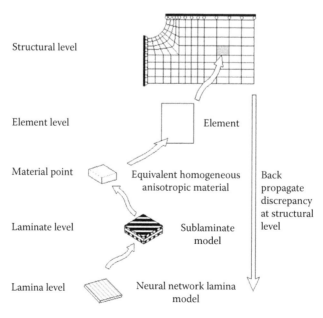

Structural level

Element level — Element

Material point — Equivalent homogeneous anisotropic material

Laminate level — Sublaminate model

Lamina level — Neural network lamina model

Back propagate discrepancy at structural level

Figure 5.9 Finite-element (FE) modeling of laminated thin composites. (From Ghaboussi, J. et al., *Int. J. Numer. Meth. Eng.*, 42, 105–126, 1998.)

5.5.4 Elastic pretraining

Elastic pretraining of the neural network material model was first carried out using a training dataset generated from the lamina elastic properties given by Chang and Lessard (1991): E1 = 22.7 × 103 ksi (156.5 GPa), E2 = 1.88 × 103 ksi (12.96 GPa), G12 = 1.01 × 103 ksi (6.96 GPa), n1 = 0.23. Radial paths in strain space were used; a total of 114 such paths that uniformly cover the strain "sphere" were selected, and training cases were generated by randomly stepping along these paths. The hidden layers each had two nodes initially; this increased to four nodes after pretraining. Thus, the pretrained neural network is described by following equation.

$$[\sigma_1, \sigma_2, \sigma_{12}] = \sigma\,NN[\varepsilon_1, \varepsilon_2, \varepsilon_{12}\,;3\,|\,2 - 4\,|\,2 - 4\,|\,3] \tag{5.28}$$

A subset of this pretraining dataset was retained in the training dataset used in the autoprogressive training that followed.

5.5.5 Autoprogressive algorithm training

Figure 5.10 shows the experimentally reported load–deflection response of the plate. Autoprogressive training was carried out for two load passes over the full range of load deflection response data. As a result of the autoprogressive training, the size of the hidden layers remained unchanged at four nodes per layer. Therefore, the trained network is described as in Equation 5.28.

After each pass, a forward analysis of the test was performed, and the results are shown in Figure 5.10. The agreement between the measured and computed responses is reasonably good at this stage but can be improved.

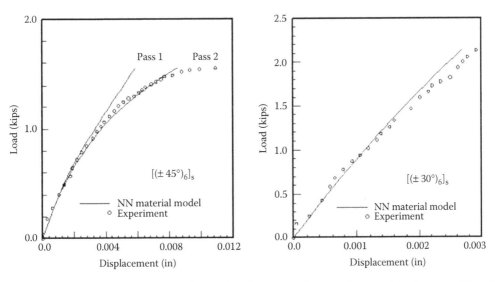

Figure 5.10 Load versus shortening from forward analysis, using the neural network lamina material model trained on [(±45°)₆]ₛ. (From Ghaboussi, J. et al., *Int. J. Numer. Meth. Eng.*, 42, 105–126, 1998.) (Experimental data from Lessard, L. and Chang, F-K., *J. Compos. Mater.*, 25, 44–64, 1991.)

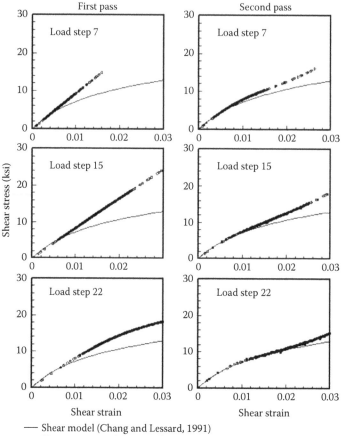

Figure 5.11 Evolution of the shear stress–strain relationship in the neural network material model. (Ghaboussi, J. et al., *Int. J. Numer. Meth. Eng.*, 42, 105–126, 1998.)

Figure 5.11 shows the shear stress and the corresponding shear strain component of the 2D training cases that are generated at different stages of the autoprogressive training during Pass 1 and Pass 2 and shows clearly how this stress–strain relationship evolves significantly with additional training. The complete 2D lamina stress–strain relationship is not fully displayed by Figure 5.11. However, the primary contributor to the nonlinearity in the structural response is the nonlinear σ_{12} versus γ_{12} relationship. An analytical shear stress–strain response model, calibrated to the test data by Chang and Lessard (1991) is also displayed in Figure 5.11 for comparison, but it is also just an approximation to the material behavior.

5.6 NONUNIFORM MATERIAL TESTS IN GEOMECHANICS

In the field of geomechanics, material tests are routinely performed on soil samples. These tests are normally used to determine some material constants used in engineering applications. They are also used to determine the material parameters for material models in FE analysis. In both applications, it is important that the state of stresses and strains within the sample to be as uniform as possible so that the sample represents a material point. Most experiments are triaxial tests where the specimen is in cylindrical shape. It is subjected to all around pressure and axial force. The friction between the specimen and the end plates at the top and bottom are eliminated as much as possible, and this leads to uniform state of stresses and strain within the sample.

Uniform material tests are fine for determining the material constants. The situation is different in training neural network material models. Neural networks learn the material behavior directly from the material tests. Uniform material tests follow a specific stress path in the stress space, and they provide information only about the material behavior along that stress path. Clearly, this is not sufficient for the neural network to learn the comprehensive material behavior to be useful in FE simulations where the stress paths followed by elements maybe unpredictable. This may require several uniform material tests, each following a different stress path so that the region of interest in the stress space is covered reasonably well. This is not a reasonable practical method; it can be expensive and it may not be practical to obtain several samples of the same material.

Autoprogressive Algorithm provides an alternative method of generating the data on the material behavior to train a neural network. We have to move away from the traditional uniform material tests to nonuniform tests where different points within the specimen follow different stress paths; the more nonuniform the better. A single nonuniform material test may provide data on the material behavior along reasonably large region within the stress space.

Professor Paul Lade performed the first nonuniform material tests on sand (Lade et al., 1994) to be used in autoprogressive training of neural network material models. Most material tests in geomechanics use triaxial test with cylindrical specimens. The specimen is subjected to axial and radial stress. To maintain a uniform state of stresses and deformations, the friction at the top and bottom of the sample is reduced as much as possible, and, if possible, it is eliminated. To generate nonuniform state of stress and deformation, Professor Lade used shorter specimens and increased the friction between the top and bottom of the sample and the end plates in the triaxial tests on sand. The friction at ends of the cylindrical specimen causes bulging around the middle that leads to nonuniform state of deformations. End frictions also cause nonuniform state of stress with shear stresses at the top and bottom of the sample.

The results of these nonuniform material tests were used in a number of studies in training neural network material models using Autoprogressive Algorithm (Sidarta and Ghaboussi, 1998; Sidarta, 2000; Hashash et al., 2009). In this section, we will present some results from the third paper. For more detailed description of the experiments and comprehensive results of the study, the reader is referred to the cited paper.

A three-dimensional FE model of the cylindrical specimen was developed using 20 node isoparametric elements. The vertical and horizontal cross-sections of the FE model are shown in Figure 5.12. The end plates, not shown in the figure, are modeled as fairly stiff blocks. The bottom nodes of the specimen are fully constrained by connecting them to the end plate, whereas the nodes at the top are radially constrained.

The material model at the integration points is represented by a Nested Adaptive Neural Network (NANN), shown in Figure 5.13. The input to the base module is the current state of stresses and strains and the strains at the end of the current increment. The output of the base module is the stresses at the end of the current increment. There are two history point modules. The inputs to the history point modules are stresses and strains at two previous increments. As described in Chapter 3, Section 3.4, the nested structure means that there are one-way connections from the history point modules to the base module. The neural network material model was initially pretrained with linear elastic material properties data generated over the $\pm 0.05\%$. The adaptive part of NANN means that during the autoprogressive training, the number of the nodes at the hidden layers may need to be increased.

The total axial load was applied in eight increments during the autoprogressive training of the neural network material model. Before starting the autoprogressive training, a forward

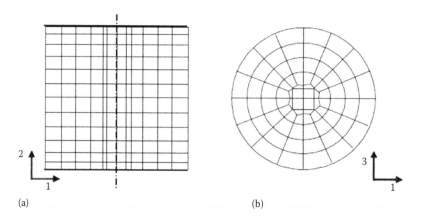

Figure 5.12 Three-dimensional FE model of the specimen in triaxial test of Ricci sand with friction end. In the FE model, the bottom is fully constrained, and the top is radially constrained: (a) vertical cross-section, (b) horizontal cross section. (From Hashash, Y.M.A. et al., *Can. Geotech. J.*, 46, 768–791, 2009.)

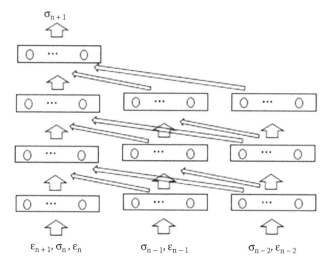

Figure 5.13 Nested adaptive neural network material model used in the autoprogressive training of the material model in nonuniform tests on Ricci sand. (From Hashash, Y.M.A. et al., *Can. Geotech. J.*, 46, 768–791, 2009.)

FE analysis was performed with the pretrained neural network; the results are shown in Figure 5.14a. Global stress–strain diagram ($\sigma_z - \sigma_r$, axial minus radial stress vs axial strain) and global volumetric strain versus axial strain are shown at the top. At the bottom, we see that lateral displacement of the specimen compared with the measured lateral displacements. Behavior of sand is highly nonlinear and because of the end restrains there are reasonably large lateral displacements. As expected, linearly elastic pretrained neural network material model is not capturing the nonlinear behavior of the system.

Autoprogressive training took place in eight passes. Trained neural network material model at the passes 4 and 8 was used in forward FE analyses, and results are shown in Figure 5.14b and c. In the first four passes, autoprogressive training was applied to only

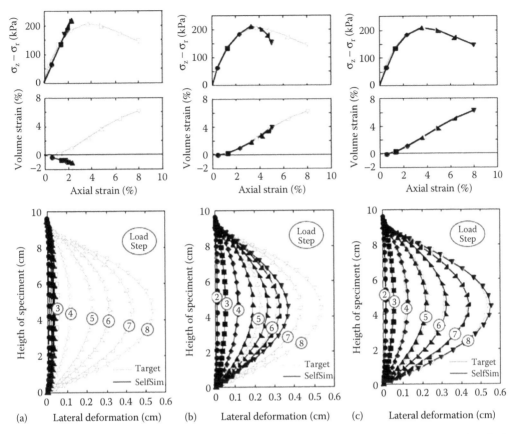

Figure 5.14 Stress–strain, volumetric strain and lateral displacement of the specimen compared with the test results at three stages of autoprogressive training of the neural network material model (From Hashash, Y.M.A. et al., *Can. Geotech. J.*, 46, 768–791, 2009.): (a) pretrained, (b) pass 4, and (c) pass 8.

the first four load increments. We can see that neural network has learned the nonlinear material behavior reasonably well for the first four increments and accurately represents the stress–strain and lateral displacement. The behavior for the load increments 5–8 is not represented well because it was not trained for those load increments.

After training for all the eight load increments in pass 8, we can see that the nonlinear material behavior is learned accurately and the FE simulation with the trained neural network represents the experimental results with a high degree of accuracy.

One of the main points about the importance of nonuniform material tests discussed earlier was that different points within the sample follow different stress paths, thus generating far more information about the material behavior than the uniform tests, where all the points within the sample follow the same global stress path. Next, we will demonstrate this point for this nonuniform material test. Figure 5.15 shows one-fourth of the vertical cross-section of the FE model of the cylindrical specimen. Vertical line on the left, labeled column A, is the vertical axis, and the horizontal line at the bottom, labeled row 1, are plane of symmetry. The points shown in this figure are the center of the FEs where stress paths are plotted in Figure 5.16.

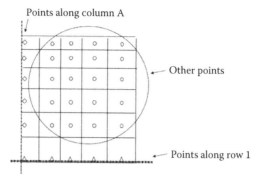

Figure 5.15 Points on one-fourth of the vertical cross-section of the FE model, where the stress paths are displayed in Figure 5.16. (From Hashash, Y.M.A. et al., *Can. Geotech. J.*, 46, 768–791, 2009.)

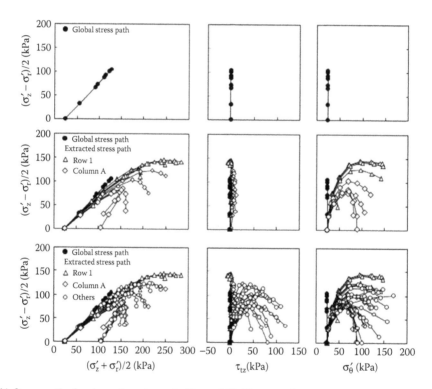

Figure 5.16 Stress paths for the points shown in Figure 5.15. Top row shows the global stress path that is the same as the stress paths followed by points in a uniform test. Rows two and three show that different points in the nonuniform test follow different stress paths that cover most of the region of interest in stress space. (From Hashash, Y.M.A. et al., *Can. Geotech. J.*, 46, 768–791, 2009.)

The top row in Figure 5.16 shows the global stress paths. In a uniform material test, all points in the sample will follow these global stress paths, thus generating limited information about the material behavior only along those stress paths.

The middle row shows the stress paths in nonuniform material test along row 1 and column A. The bottom row shows the stress paths for all the points indicated in Figure 5.15. We can clearly see that these stress paths cover a large part of the stress space and thus contain far more information about the material behavior than along the global stress paths. Of course, we have seen that Autoprogressive Algorithm makes it possible to extract the information about material behavior along these stress paths. Without Autoprogressive Algorithm non-uniform test results will be practically useless for training a neural network material model.

5.7 AUTOPROGRESSIVE TRAINING OF RATE-DEPENDENT MATERIAL BEHAVIOR

Up to this point, we have discussed application of Autoprogressive Algorithm in training of neural networks to learn the rate-independent material behavior. Autoprogressive Algorithm has also been successfully used in modeling of the rate-dependent constitutive behavior of materials (Jung, 2004, Jung and Ghaboussi, 2006a, 2006b, 2010, Jung et al., 2007). In this section, we will briefly describe the application of Autoprogressive Algorithm in modeling of the rate-dependent constitutive behavior of materials and present a simple illustrative example. More detailed information is available in the references cited above.

Neural networks can be used to model the viscoelastic as well as the nonlinear rate-dependent material behavior. In the case of isotropic materials, the rate-dependent properties of most materials can be divided into volumetric and deviatoric parts. Volumetric stresses and strains (σ_v or p = pressure, ε_v) and deviatoric stresses and strains (s_{ij}, e_{ij}) are defined in the following equations:

$$\begin{cases} \sigma_v = p = \frac{1}{3}\sigma_{kk} \\ \varepsilon_v = \varepsilon_{kk} \\ s_{ij} = \sigma_{ij} - \delta_{ij}\sigma_v \\ e_{ij} = \varepsilon_{ij} - \frac{1}{3}\delta_{ij}\varepsilon_v \end{cases} \tag{5.29}$$

Since in isotropic materials the volumetric and deviatoric material properties can be separated, two separate neural networks can be used to model them. Typically, these two neural networks can be represented as follows:

$$\begin{cases} \dot{\sigma}_v^n = NN\left[\sigma_v^{n-1}, \varepsilon_v^{n-1}, \sigma_v^n, \varepsilon_v^n, \dot{\sigma}_v^{n-1}, \dot{\varepsilon}_v^{n-1}, \dot{\varepsilon}_v^n; \quad \right] \\ \dot{s}_{ij}^n = NN\left[s_{ij}^{n-1}, e_{ij}^{n-1}, s_{ij}^n, e_{ij}^n, \dot{s}_{ij}^{n-1}, \dot{e}_{ij}^{n-1}, \dot{e}_{ij}^n; \quad \right] \end{cases} \tag{5.30}$$

The neural network architecture in this equation is left black since it represents a general neural network material model.

These two neural networks can be trained in Autoprogressive Algorithm. We have seen that the neural networks need to be pretrained before the start of Autoprogressive Algorithm. If some training data are available that is close to the expected material behavior of the system, it can be used in pretraining of the neural material models. Otherwise, approximate

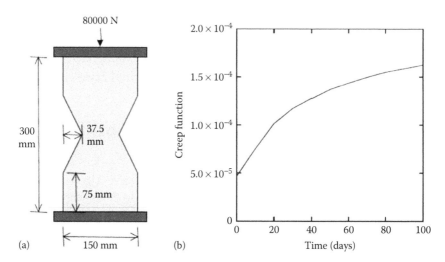

Figure 5.17 The specimen and material properties of the simulated experiment: (a) cylindrical specimen in the simulated experiment and (b) rate-dependent material used in the simulated experiment. (From Jung, S-M. and Ghaboussi, J., *Comput. Methods Appl. Mech. Eng.*, 196, 608–619, 2006b.)

linear elastic data are generated and used in pretraining of rate-independent material model. The situation is somewhat different in the rate-dependent material properties. Since the rate derivatives of stresses and strains are involved, linear elastic material properties are not sufficient in pretraining; a viscoelastic material model is needed to pretrain the neural network material models. Next, we will present a simple illustrative example.

The system used in the simulated experiment, shown in Figure 5.17a, is cylindrical solid with variable cross-section to induce nonuniform state of stresses and strains. It is assumed that there is no friction between the specimen and the rigid end plates. The volumetric component of the material properties is assumed to be linearly elastic and rate independent. The deviatoric creep function (e/s vs. time) is shown in Figure 5.17b. It is assumed that the deviatoric components have the same rate-dependent material properties, and the following neural network is used to represent it.

$$\dot{s}^n = NN\left[s^{n-1},\ e^{n-1},\ s^n,\ e^n,\ \dot{s}^{n-1},\ \dot{e}^{n-1},\ \dot{e}^n;\qquad \right] \tag{5.31}$$

The material properties used in the simulated experiments are given in Jung and Ghaboussi (2006b).

The specimen is modeled with 256 axisymmetric four-node isoparametric elements, and it is subjected to a constant axial force and the vertical displacement is monitored. The results of the simulated test are used in the Autoprogressive Algorithm. The neural network material properties are initially pretrained with the data generated from a viscoelastic material model with material properties that were different than the material properties used in the simulated experiment. In the Autoprogressive Algorithm FEA1, the specimen was subjected to the applied axial force, and in the FEA2 it was subjected to the measured axial displacement at the top. The load was applied in 10 increments, and iterations were performed in each increment to satisfy the specified convergence criterion.

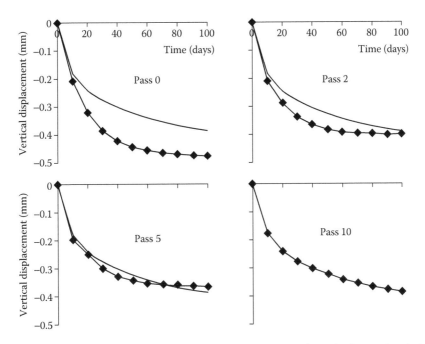

Figure 5.18 Comparison of the displacement at the top of the specimen from the forward analysis with the neural network material model at the pretrained and passes 2, 5, and 10 with the results of the simulated experiment. (From Jung, S-M. and Ghaboussi, J., *Comput. Methods Appl. Mech. Eng.*, 196, 608–619, 2006b.)

Ten passes of Autoprogressive Algorithm were completed. The neural network at pretrained state and at the end of passes 2, 5, and 10 were used in forward FE analysis. The displacements at the top are compared with the results of the simulated experiment in Figures 5.18.

Figure 5.19 shows the strains at the point shown on the top of the figure, from forward analyses at pretrained state and at the end of passes 5 and 10. The strains refer to the axi-symmetric coordinate system (r, θ, z). The results from Figures 5.18 and 5.19 are typical of the behavior at the other points in the specimen. These results clearly demonstrate that the neural network material model has been able to learn the rate-dependent material properties in Autoprogressive Algorithm.

This method has also been applied in determining the rate-dependent material properties of concrete from the measured response of the reinforced concrete structures (Jung and Ghaboussi, 2010; Jung et al., 2007). This has potential application in construction of certain types of bridges. Rate-dependent properties of concrete cannot be accurately determined from laboratory tests; they are also affected by the environmental conditions, such as temperature and humidity. With the application of Autoprogressive Algorithm, it is possible to accurately determine and calibrate the rate-dependent material properties of the concrete in the early stages of construction and use that information to forecast the response the structure in the later stages of construction. This will allow one to take timely remedial action. This was demonstrated by applying Autoprogressive Algorithm to a segmental rein-forced concrete bridge (Jung et al., 2007).

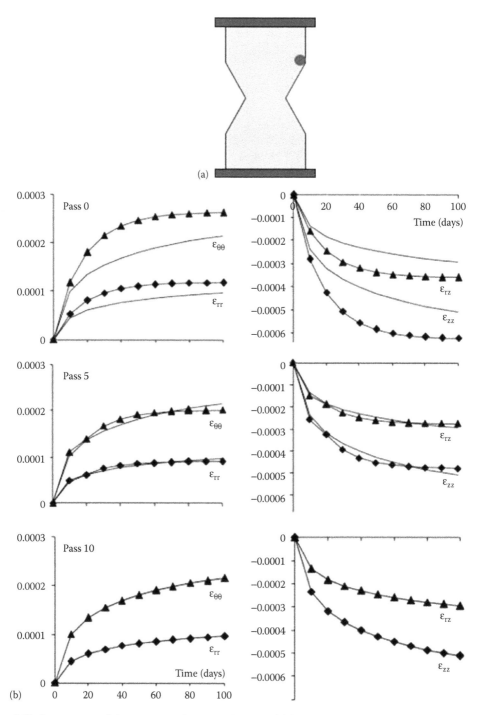

Figure 5.19 Comparison of the strains at the point shown in (a) from the forward analysis with the neural network material model at the pretrained and passes 5 and 10 with the results of the simulated experiment (From Jung, S-M. and Ghaboussi, J., *Comput. Methods Appl. Mech. Eng.*, 196, 608–619, 2006b.): (a) the point at which the strains are shown below and (b) strains the point shown above.

5.8 AUTOPROGRESSIVE ALGORITHM IN BIOMEDICINE

Autoprogressive Algorithm has potential in determining the mechanical properties of the soft tissue in biomedical imaging. Most imaging methods create a qualitative image of some property of the soft tissue, without determining the actual value of those properties. Autoprogressive Algorithm has the potential of creating quantitative images of soft tissue that may contain important medical information potentially useful in diagnosis and treatment. We will briefly describe research on application of Autoprogressive Algorithm in two areas: in ophthalmology; and in elastography. These two applications of the Autoprogressive Algorithm are presented as illustrative examples to indicate much broader potential applications in biomedical imaging.

The first example (Kwon, 2006; Kwon et al., 2008, 2010; Ghaboussi et al., 2009) is on using soft computing methods to address an important issue in the field of ophthalmology; accurate determination of intraocular pressure (IOP) using Goldman's Applanation Tonometer (GAT). IOP is the pressure of aqueous humor in the anterior chamber (between cornea and lens). Applanation tonometer is pressed against the cornea until the area under it is flat. The measured force divided by the flattened area of cornea is used as an approximation of the IOP. An FE model of the human cornea subjected to GAT is shown in Figure 5.20.

For thin human corneas, the measured IOP is fairly accurate and close to the actual IOP. The problem arises with the thicker human corneas where Goldman Applanation Tonometer can highly overestimate the IOP. The reason is that part of the applied force must overcome the bending resistance of the cornea that are higher in the thicker corneas. Using the total force in determining the IOP leads to much higher values than the true IOP that can lead to unnecessary treatment.

Accurate determination of IOP from GAT requires simulating the process of applanation, as shown in Figure 5.20, that requires determination of mechanical properties of human cornea that may also have other applications beyond the accurate determination of IOP; for example, in simulating laser surgery a priori, in order to optimize the outcome (Kwon, 2006). We note that the human cornea material properties are nonlinear and anisotropic. This is a difficult type-two inverse problem. Two approaches were used in addressing this problem.

First, a combination of genetic algorithm and neural network (GA/NN) were used to determine the IOP and the parameters of a material property model for the human cornea (Ghaboussi et al., 2009). This method will be briefly described in Chapter 6.

The second approach uses a two-step method to determine a more accurate neural network material model of the human cornea. In the first step, a combination of the GA/NNs is used

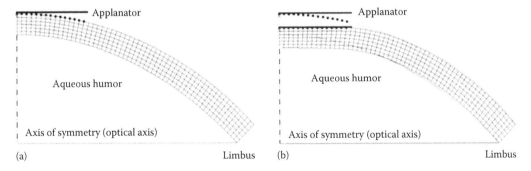

Figure 5.20 FE model of human cornea subjected to applanation tonometer: (a) axisymmetric finite element model of the cornea and (b) cornea surface flattened by applanator. (From Kwon, T-H., Minimally invasive characterization and intraocular pressure measurement via numerical simulation of human cornea, PhD thesis, Department of Civil and Environmental Engineering, University of Illinois at Urbana-Champaign, Urbana, IL, 2006.)

to determine the accurate value of the IOP. The computed IOP is used in the Autoprogressive Algorithm to determine the neural network material model of the human cornea (Kwon et al., 2008, 2010).

In the second application, Autoprogressive Algorithm is applied to quasi-static ultrasonic imaging to determine the mechanical properties of soft tissue. This general area is also referred to as *elastography* in biomedical imaging.

Mechanical properties of soft tissues can provide information about its local health status. The cells in pathological tissues may form stiff extracellular regions, which can be early diagnostic indicator of disease. Quasi-static ultrasonic imaging provides an image of the subsurface strains that give an approximate indication of relative stiffness with some diagnostic value. Accurate determination of mechanical properties of soft tissue also requires image of the stresses that is not directly provided by the quasi-static ultrasonic imaging. With the surface force and volumetric displacement data gathered during imaging, we have a type-two inverse problem of determining the mechanical properties of the soft tissue. Autoprogressive Algorithm is used to solve this problem. Some early results are reported in Hoerig et al. (2016, 2017).

5.9 MODELING COMPONENTS OF STRUCTURAL SYSTEMS

Up to this point in this chapter, we have mainly concentrated on application of Autoprogressive Algorithm in determining the constitutive models that define the material properties at a single point. Autoprogressive Algorithm has broader potential applications. It can also be applied to model the behavior of regions or components of structural systems. As an example, we will consider the beam–column connections in moment resisting steel frames of structural systems.

The behavior of bolted and/or welded beam-to-column connections in steel frames has an important effect on their structural response under earthquake ground motion. Their hysteretic response exhibits highly inelastic characteristics and continuous variation in stiffness, strength, and ductility. This is due to yielding of materials, buckling of components, frictional slippage between components, fracture, and slacking of fastened bolts. Connecting elements, including angles, plates, and T-studs, often experience yielding mechanism or local failure and the interactions between the components heavily influence the cyclic behavior of the connection. Hysteretic behavior of four typical beam–column connections is shown in Figure 5.21.

Accurate hysteretic models of bolted connections are important in modeling and analysis of structural systems for seismic assessment and design. Beam-to-column joints have served as energy dissipation regions under extreme dynamic loads such as earthquakes. Experimental and analytical studies have been undertaken to investigate the feasibility of using bolted connections in steel frames as points where seismic energy is absorbed through hysteretic response that increases their seismic stability. Although welded connections are traditionally used for seismic designs, past research has shown that semirigid connections may be effectively used for seismic design and the large spectrum of their behavior strongly influences frame stability and strength. To take advantage of semirigid connections, it is necessary to represent the actual joint hysteretic behavior with reasonable accuracy in assessment and computational simulation for design.

Accurate modeling of the hysteretic behavior of welded and/or bolted beam-to-column connections require detailed FE modeling, including the nonlinear material properties, regional and/or component buckling and/or fracturing, large deformations, and slipping at frictional contacts. This type of detailed FE is suitable in studying the hysteretic behavior of the connections. However, these detailed models that may require hundreds or thousands of FEs cannot be used effectively in dynamic analysis of the whole frames where the

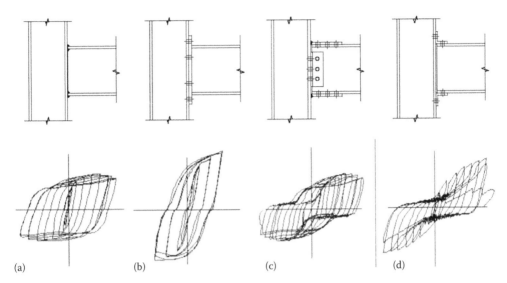

Figure 5.21 Four steel beam–column connections and their hysteretic moment-rotation behavior at the end of the beam: (a) fully welded connection, (b) end-plate connection, (c) flange–plate connection, and (d) angle connection. (From Kim, J.-H., Hybrid physical and informational modeling of beam-column connections, PhD thesis, Department of Civil and Environmental Engineering, University of Illinois at Urbana-Champaign, Urbana, IL, 2009.)

beam–column connections are modeled as points at the ends of beams. These are the locations where the model of the hysteretic moment-rotational response is needed.

There are simplified mechanical models of the beam–column connections that approximate their hysteretic behavior (Elnashai et al., 1998). These models tend to approximate the main features of the hysteretic behavior of beam–column connections but lack the ability to capture some more complex aspects of the behavior.

Neural network hysteretic behavior of the point models of the connections can be developed in two ways: (1) training the neural network directly with the training data that is acquired from the experiments or from the detailed FE simulation (Yun et al., 2007; Kim, 2009; Kim et al., 2010); (2) using Autoprogressive Algorithm to train the neural network model of the beam–column connection from the measured response of the steel frame (Yun et al., 2007, 2008b).

In both approaches, the neural network is trained to model the complete moment-rotation response of the connection. Next, we will briefly describe a method using the neural network to complement a mechanical model of the connection: a *hybrid mechanical–informational model.*

5.10 HYBRID MATHEMATICAL–INFORMATIONAL MODELS

We have seen that neural networks are capable of learning constitutive behavior of materials as well as behavior of components of structural system. We have also seen that these neural network models can either be trained directly when the appropriate data are available, or they can be trained by solving the inverse problem through Autoprogressive Algorithm and self-learning simulation. It is also possible to train neural networks to learn the behavior of structural system itself, although we have not given any examples for that. We can refer to these neural network models as informational models; they are trained with data that contain the appropriate information.

In most cases, there are mathematical models that describe the behavior of materials, components of structural systems and whole structural systems. Obviously, these mathematical models are based on mathematics, physics, and mechanics. There are no reasons to only deal either with mathematical or informational models. We can also have hybrid models. We start with a mathematical model that is often based on a simplified mechanical model. Neural networks are used to capture some aspects of the behavior that the simplified mechanical models miss.

This is schematically illustrated in Figure 5.22. On the right, there are the mathematical models, and on the left, there are the informational models. The portion of each modeling approach decreases as we move towards the other modeling approach whose contribution increases. Between the two basic modeling approaches is the range of the hybrid mathematical–informational models. We will illustrate this approach with an example of the beam–column connection (Kim, 2009; Kim et al., 2010, 2012). It is important to point out that hybrid mathematical and informational models are made possible by Autoprogressive Algorithm and self-learning simulation, as will become clear in the following example.

A bolted flange–plate connection is shown in Figure 5.23. This is one of eight full-scale bolted flange–plate connections that were tested in the Newmark Structural Engineering Laboratory at the University of Illinois at Urbana—Champaign (Schneider and Teeraparbwong, 2002); the one used in this study was labeled BFP06. Also shown in this figure is the simplified mechanical model of the flange plates and column panel zone. These simplified mechanical models are intended for use in the dynamic analysis of the steel frames subjected to earthquake ground motion. More detailed FE models of the connections

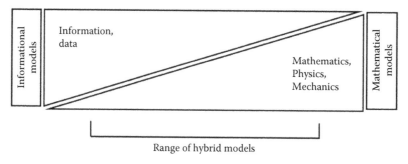

Figure 5.22 Schematics of hybrid models that supplement the mathematical models with neural network based informational models.

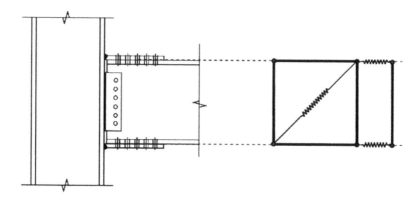

Figure 5.23 A bolted flange–plate connection and its mechanical model. (From Kim, J-H. et al., Eng. Struct., 45, 1–11, 2012.)

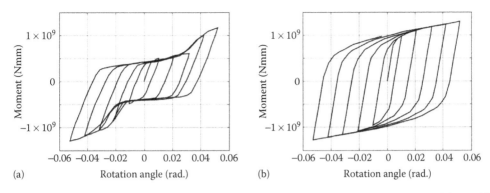

Figure 5.24 Hysteretic behavior of bolted flange–plate connection from (a) experiment and (b) simplified mechanical model. (From Kim, J-H. et al., *Eng. Struct.*, 45, 1–11, 2012.)

that include nonlinearity, yielding, slippage, and fracture are possible but not practical for use in dynamic analysis of the steel frames.

The experimentally measured hysteretic moment-rotation behavior for the bolted flange–plate connection is shown in Figure 5.24. Also shown in this figure is the simulated hysteretic response of the simplified mechanical model. Experimental results show significant amount of pinching in the hysteretic behavior of the connection. The simplified mechanical model lacks the aspects of the connection that contribute to the pinching effect. It is reasonable to observe that the slippage and yielding of the flange–plates and the panel zone distortion are the major sources of inelastic deformation that contributes to the inelastic behavior and pinching. These details are not included in the simplified mechanical models.

In a hybrid model, the informational part can capture the aspects of the hysteretic behavior that simplified mechanical model misses, such as factors that contribute to the pinching. In this hybrid modeling, the slippage between the flange and plate is the main component of the hysteretic behavior to be modeled with neural networks. The hybrid model is presented in Figure 5.25. The neural network in the hybrid model is trained with Autoprogressive Algorithm. Hybrid model can also be at several levels. In this case, we have two levels. In level one, NN1 is trained first with the Autoprogressive Algorithm from the experimental results. The data for training NN2 can be extracted by using a second-level Autoprogressive Algorithm. In this example, the linear decomposition technique was used for the second level instead of Autoprogressive Algorithm. Finally, force–displacement pairs are collected for the

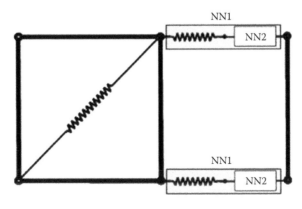

Figure 5.25 Hybrid Mechanical and informational model of the bolted flange–plate connection. (From Kim, J-H. et al., *Eng. Struct.*, 45, 1–11, 2012.)

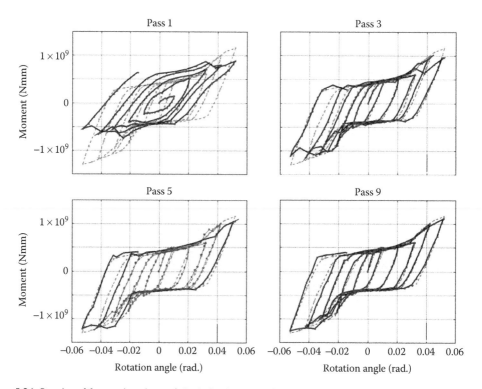

Figure 5.26 Results of forward analysis of the hybrid model of the bolted flange–plate beam–column connection at four passes during the Autoprogressive Algorithm training (solid lines) compared with the experimental results (dashed lines). (From Kim, J-H. et al., *Eng. Struct.*, 45, 1–11, 2012.)

training data of the target component. The final neural network NN2 is given by the following equation that relates displacement d to force f:

$$f_n = F \, NN[d_n, \, d_{n-1}, \, f_{n-1}, \, \xi_n, \, \Delta\eta_n; \, 5 \mid 20 \mid 20 \mid 1]$$

$$\begin{cases} \xi_n = f_{n-1} \, d_{n-1} \\ \Delta\eta_n = f_{n-1} \, \Delta d_n \end{cases} \tag{5.32}$$

This form of neural network material model for cyclic behavior was discussed in Chapter 3, Section 3.7.

The neural network material model NN2 was initially pretrained with linear elastic material properties over a small range of the displacements around the origin. The results of the Autoprogressive training in four passes are shown in Figure 5.26. At the end of each pass, the hybrid model with the neural network NN2 was subjected to cyclic moments and the computed behavior of the hybrid model is compared to the experimental results.

It is clear from this figure that the neural network in the hybrid model has learned to compensate for the aspects of the hysteretic behavior of the connection that the simplified mechanical model was not capable of representing.

This example clearly demonstrates the power of the hybrid models trained in Autoprogressive Algorithm and self-learning simulations to learn reasonably complex hysteretic behavior of beam–column connections. It is important to point out that the hybrid modeling has much broader potential applications to wide range of complex problems in mechanics and beyond.

Chapter 6

Evolutionary models

6.1 INTRODUCTION

In earlier chapters, we discussed the trend toward biologically inspired models. To reiterate briefly, we have seen that some biological systems have certain capabilities that either do not exist in our engineered world or they are far superior to the equivalent capabilities in our engineered world. Biological systems seem to exploit the power of the massively parallel systems of inter related agents, combined with the power of random search. In the previous chapters, we considered neural networks that are systems of massively interconnected artificial neurons. In this and the next two chapters, we will consider models that are based on the natural Darwinian evolution. As we will see later, these models are also massively parallel systems, in that they deal with large populations of individuals who are interrelated by competing within the same shared environment. Evolutionary models exploit the power of random searches by introducing random variability at every stage of the evolutionary computation.

Darwinian evolution offers a clear explanation for emergence of higher life forms in nature. Evolution is highly complex, but the basic elements of the theory of evolution are simple. Darwinian evolution is based on the fact that competition for survival, after many generations, favors the survival of the fittest individuals, thereby giving them a better chance of producing offsprings and passing their superior genetic code to the next generations. This process will gradually lead to fitter individuals who are more adapted to their environment. This process, which is also called natural selection, requires some random changes from one generation to the next, which is provided by recombination and mutation.

On a very fundamental level, a process that results in highly complex living systems, far more complex than anything we can engineer, should also be considered a potential engineering design tool. Progress in computing has made this possible. The first evolutionary-based model was genetic algorithm, which was developed by John Holland at University of Michigan in the 1960s (Holland, 1975).

Genetic Algorithm has been mainly considered and used as an optimization method by most researcher and practitioners. A great advantage of the genetic algorithm is that, unlike some of the classical optimization methods, it only requires the value of the fitness function, not its gradient that is often very difficult, if not impossible, to determine. The fitness function in genetic algorithm can be made as complex as possible, as long as its value can be determined. As we will see later, in those cases where the fitness evaluation requires a computer simulation, the most computationally intensive part of the analysis with genetic algorithm can be the fitness evaluation. Although certain problems can theoretically be tackled with the genetic algorithm, the prohibitive cost of the function evaluation may put them out of the reach of most practical applications.

As a method based on the natural evolution, genetic algorithm has considerable potential far beyond the simple optimization. As we will see later in Chapters 7 and 8, it can be developed into a design method that may also include optimization as an integral part of the process. This is an important point. To further explore this point, we need to examine the fundamental differences between engineering optimization methods and the process of evolution.

Mathematical optimization methods, such as dynamic programing, aim at determining the mathematically exact optimum solution. The intermediate steps on the away to the mathematically exact optimum solution may not contain any useful information and may not result in a sequence of improving solutions. It is important to note that at least one optimum solution must exist for the optimization methods to succeed, which in turn requires that the solution space be finite dimensional. Of course, in a finite-dimensional solution space, many local optimum solutions may exist.

Evolutionary based models use a completely different strategy. To understand the basis of these methods, it is important to examine the process of natural evolution. We have discussed the fact that evolution leads to fitter organisms who are better adapted to the environment. Evolution uses the gradual improvement to create fitter and more adapted organisms, and it always uses the current information. Evolution does not seek a mathematically optimum, because it always operates in an open-ended infinite-dimensional solution space that does not contain any optimum solutions. The process of improvement can continue forever, especially under changing environment.

To further explore this concept let us ponder the following question: Are the human beings optimal? After careful consideration of this question we are likely to arrive at the conclusion that this question does not have an answer. The reason is that there is no optimal solution in the solution space in which we have evolved. It is an infinite-dimensional solution space. Our current form is the consequence of our evolutionary past and the specific route that it has taken through the solution space of all the possible forms. If the environment remains stable, then the organisms living in that environment may approach a more or less stable configuration. In such a case, it may be tempting to speak of moving toward the optimum forms. However, if the environment changes, then the evolution may take an unexpected path through the infinite-dimensional solution space.

In a very restricted and hypothetical sense, we may also think of evolution as operating in a finite-dimensional solution space. This hypothetical situation arises when we have a single specie evolving in a fixed and unchanging environment, and only a finite number of the attributes of the individuals may change. Majority of evolutionary-based models are based on this highly idealized version of the natural evolution. In these finite-dimensional solution spaces, there are at least one and possibly more local optimum solutions. Where there is an optimum solution, there are also many near optimal solutions that are arbitrarily close to the mathematically exact optimum. While the mathematically based optimization methods seek the exact optimum, genetic algorithm approaches that optimal solution through continuous improvement. The returns become diminishingly small as we approach the exact optimum solution. Similar to the natural evolution, there is no incentive to go beyond the near optimal solution. The basic strategy of the evolutionary models is to use improvement based on the current information to reach a near optimal solution in a potentially open-ended solution space.

We will see later that the discussion of the evolution in infinite-dimensional solution space versus the finite-dimensional solution space is not purely theoretical. In Chapter 9, we will discuss the applications of genetic algorithm in design, particularly engineering design, which may approach the conditions similar to those of a search in an infinite-dimensional solution space, whereas most optimization problems are searches in a finite dimensional solution space.

6.2 EVOLUTION AND ADAPTATION

Evolution produces increasingly fit organisms who adapt to their environment in the presence of high degree of uncertainty, chance, and variability. Fitness and adaptation is a measure of the performance of the organism in its environment compared with the other members of the population. An individual organism's chance of survival in the nature is fairly uncertain no matter how fit it may be. However, in a large population, a sufficient number of organisms who have adapted to their environment more than the others have a better chance of survival to pass their genes with more adapted traits to the next generation. Adaptation only occurs in a population, and it is the result of the interaction of an organism both with its environment as well as with the other members of the population. The latter is competitive as well as cooperative, competition for the resources needed for survival and cooperation in mating and producing offsprings.

Competition is of course the key for evolution and adaptation. Although it is difficult to think of natural environment where competition does not take place, it is possible to do so in artificial environments. Some form of evolution does also occur in the industry and business. For example, consider the cars. They have been going through significant changes since their introduction and mass production. Many of the improvements have been the result of the competition between the car manufacturers, and this process still continues. Similar situations do also exist in other industries, especially young industries. It is also known that when competition ceases, the pace of innovation can slow down considerably, and it may even come to a stop. A good example of this is when monopolies develop.

An important feature of adaptation is that it occurs gradually and progressively through modification of some internal structure, which is often random in nature. Competition occurs at a given time and between organisms that are at the same evolutionary stage. At each stage of evolution, the organism performs well enough to survive in their environment. The evolutionary process produces useful solutions along the way. It is also a fact that evolution is a slow and very long process. In fact, it is also an open-ended process.

The evolutionary process and its workings are highly complex. However, the basic components of evolutionary process are listed below.

Inheritance and recombination of the genetic codes: The genetic codes pass from one generation to the next. The offsprings inherit their genes from their parents. Some form of recombination of the genes of the parents does occur in process.

Random mutations: The offsprings do not inherit the exact copy of the genes from their parents. Some random changes do occur in passing the genetic information from parents to the offsprings. These changes may result in higher or lower fitness.

Competition for survival in an environment: Organisms must compete for resources needed to survive. Fitter individuals have a higher chance of survival and passing of their genetic codes to the next generation. As a result, the next generation inherits the genetic characteristic more suitable for survival. This leads to gradual increase in fitness and adaptation.

Increased fitness and adaptation of the individuals to their environment: The competition and survival of the fitter member gradually and progressively lead to higher levels of fitness and adaptation.

6.3 GENETIC ALGORITHM

In this section, we will describe the Simple Genetic Algorithm (SGA) and present the basic elements of this class of the evolutionary based methods. It is not implied that SGA is the best nor the most effective. Researchers have proposed many variations of the basic genetic algorithm method. For the purposes of this book, the understanding of the basic element of the method is of paramount importance for the topics to be discussed in the next two chapters.

A computational evolutionary adaptive model will have some of the important basic elements of evolution, albeit in a much simplified and idealized form.

6.3.1 Population of genetic codes

Genetic algorithm tries to imitate the evolutionary process in computational simulation for solving an optimization or an optimal design problem. The members of the population are simply genetic string that encodes the problem to be optimized. The encoded information includes the elements of the problems that are to change during the optimization process and it may include the variables and parameters of the problem. There are many ways of encoding information in a genetic code. The most common form is binary encoding. A string of binary numbers constitutes the genetic code. It is usually parsed from left to right and each predetermined sequence of binary bit represents one number. For example, a population of four binary strings are shown in Figure 6.1, each have 40 binary bits. Each of these strings could represent five numbers of 8 bits each.

The genetic code for even small practical problems may contain several thousand binary bits. Also, the number of bits required for different variables and parameters may vary. The same population may also be integer encoded. The following figure shows the integer encoded population. The variables and their values in the binary encoded and integer encoded strings are identical (Figure 6.2).

We will denote the variable field encoded in the binary strings by **S**.

$$S = \left[x_1, \cdots, x_N \right] \tag{6.1}$$

The N variables are denoted by x_1 through x_N.

[00101110][01001111][01001111][00101011][10101010]
[00010010][00001100][00001100][00100110][10100101]
[00101100][11010001][11010001][01010110][01010001]
[00101001][00000111][00000111][00011100][01110100]

Figure 6.1 A population of fixed length binary encoded string genotypes with five variables.

[92][141][157][89][340]
[36][89][24][79][325]
[88][145][417][172][161]
[81][72][13][56][232]

Figure 6.2 A population of four fixed length integer encoded string genotypes with five variables.

As we discussed earlier, evolution and adaptation takes place in population of individuals collectively interacting with their environments. Similarly, in genetic algorithm, a population of strings are needed. The size of the population has to be large enough to allow sufficient amount of variability. The minimum population size is highly problem dependent. Population sizes of anywhere from 50 and to many thousands have been used. It is surprising that significant evolution can occur even in a small population. We will present an illustrative example with a population of only 30.

6.3.2 Artificial environment and fitness

In genetic algorithm, the environment is usually not defined explicitly. The fitness f(S) is defined explicitly as a measure of relative performance in the hypothetical environment. The fitness function can be defined in many ways. It is defined such that the value of the function increases as the state S approaches the optimum solution. The performance of the genetic algorithm in terms of the rate of convergence may depend on the form of the fitness function. However, here again there is no one rule that may be applicable in all cases and the situation is highly problem dependent.

6.3.3 Competitive rules of reproduction and recombination

The key operator in genetic algorithm is the rules of reproduction and recombination. The two main parts are the method of selecting the mating pairs and a method of recombining their genetic codes to produce the members of the next generation. These rules should embody the principle of competition in an environment with a degree of uncertainty such that the fitter members should have a better chance of being selected as the mating pairs. Again, there are many methods of achieving these goals, and they mainly differ in detail and degrees. The simplest and the most common rule is called the "roulette wheel method." Assume that the circumference of a hypothetical roulette wheel is assigned to each individual in the population according to its fitness. The fitter individuals have a larger portion of the roulette wheel than the less fit individuals. We can also demonstrate this in a linear form that is perhaps easier to visualize. Consider a population of N individuals with fitness function values of f_1 through f_N. An additional step is to determine the normalized fitness functions.

$$\bar{f}_i = \frac{f_i}{\sum\limits_{j=1}^{N} f_j} \quad i = 1, \dots N \tag{6.2}$$

A linear segment is then allocated according to the normalized fitness functions. The selection of the mating pairs takes place by choosing a random number in the interval [1, N]. This is illustrated in Figure 6.3 for a population of eight individuals. In this case, N = 8 and two random number of 1.2 and 5.7 would select the string numbers 2 and 7 as the mating pair. This process may also result in the selection of the same member as both parents. In that case, the crossover operator will not have any effect and two identical copies of the parent string will be produced in the next generation. It is for this reason that the selection of the same member as both parents may be prevented. The weighted random process of selection of the mating pairs is continued until all the parents of the next generation are selected.

As for the recombination, again there are many rules. The simplest is the method originally proposed by John Holland and is still the most frequently used. This method is called the crossover rule. A crossover location is randomly selected along the string and the portions the strings of the mating pair beyond that point are switched, as shown in Figure 6.4. This is a one-point crossover. Multiple-point crossovers can be used for longer strings.

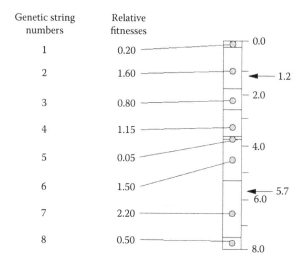

Figure 6.3 Schematics of the random selection of the mating pairs. Each string's chance of being selected is proportional to its relative fitness.

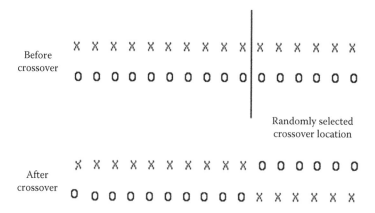

Figure 6.4 Schematic illustration of crossover of a pair of binary-encoded strings.

6.3.4 Random mutation

This is the mechanism of introducing random changes in the population at a slow rate. Usually, a rate of mutations of 0.01 or 0.001, or a similar small number, is used. Mutation rate depends on the population size and the string lengths. Bits from the genetic codes are selected at the mutation rate and switched. In cases where the genetic codes are encoded by other than the binary bits, the values are simply replaced by another random number.

6.3.5 Illustrative example

To clarify the basic operators of the Genetic Algorithm, we examine the details of transition from one generation to the next in a simple problem. The problem considered involves four variables, x_1 to x_4, and the fitness function is defined as the product of those variables. The fitness function for the jth string is given in the following equation:

$$F_j(x_i) = \prod x_i \qquad (6.3)$$

x_1	x_2	x_3	x_4	Fitness, F
[00101110]	[01000111]	[01001111]	[00101011]	11,094,602
[00010010]	[00101101]	[00001100]	[00100110]	369,360
[00101100]	[01001001]	[11010001]	[01010110]	57,732,488
[00101001]	[00100100]	[00000111]	[00011100]	289,296
[10101010]	[10100101]	[01010001]	[01110100]	263,557,800
[10010010]	[11010010]	[01100100]	[00010100]	61,320,000

Figure 6.5 Illustrative example: a population of six binary encoded strings before crossover. The last column shows the fitness of the strings, evaluated as the product of the numerical values of the variables.

The variables are encoded by eight binary bits each in a population of six individuals, as shown in Figure 6.5. The last column shows the fitness values, which are calculated by converting the binary strings to integer number. For example, the value of x_1 in the first string is calculated as follows:

$$x_1 = 0\times128 + 0\times64 + 1\times32 + 0\times16 + 1\times8 + 1\times4 + 1\times2 + 0\times1 = 46$$

The calculations of the values of the fitness function are shown below:

$$F_1 = 46\times71\times79\times43 = 11,094,60$$
$$F_2 = 18\times45\times12\times38 = 369,360$$
$$F_3 = 44\times73\times209\times86 = 57,732,488$$
$$F_4 = 41\times36\times7\times28 = 289,296$$
$$F_5 = 170\times165\times81\times116 = 263,557,800$$
$$F_6 = 146\times210\times100\times20 = 61,320,000$$

Let us assume that a random selection weighted by the fitness function value has resulted in three pairs: strings 3 and 6; strings 1 and 5; strings 4 and 5. The crossover locations have been randomly determined to fall after the bits 17, 28, and 8 for the three pairs. The crossover operation is shown in Figure 6.6.

String pairs 3 and 6 and crossover point after bit number 17

[00101100] [01001001] [1   1010001] [01010110]
[10010010] [11010010] [0   1100100] [00010100]

String pairs 1 and 5 and crossover point after bit number 28

[00101110] [01000111] [01001111] [0010   1011]
[10101010] [10100101] [01010001] [0111   0100]

String pairs 4 and 5 and crossover point after bit number 8

[00101001]   [00100100] [00000111] [00011100]
[10101010]   [10100101] [01010001] [01110100]

Figure 6.6 Illustrative example: crossover of the randomly selected pairs.

$$x_1 \qquad x_2 \qquad x_3 \qquad x_4$$

[00101100][01001001][11100100][00010100]
[10010010][11010010][01010001][01010110]
[00101110][01000111][01001111][00100100]
[10101010][10100101][01010001][01111011]
[00101001][10100101][01010001][01110100]
[10101010][00100100][00000111][00011100]

Figure 6.7 Illustrative example: population of the next generation of binary encoded strings after crossover.

x_1	x_2	x_3	x_4	Fitness, F
[0 0 1 0 1 1 0 0]	[0 1 0 0 1 0 0 0)]	[1 1 1 0 0 1 0 0]	[0 0 0 1 0 1 0 0]	14,446,080
[1 0 0 1 0 0 1 0]	[1 1 0 1 0 0 1 0]	[0 1 0 1 0 0 0 1]	[0 1 0 1 0 1 1 0]	213,557,560
[0 0 1 0 1 1 1 0]	[0 1 0 0 0 1 1 1]	[0 1 0 0 1 1 1 1]	[0 0 1 0 0 1 0 0]	9,288,504
[1 (1) 1 0 1 0 1 0]	[1 0 1 0 0 1 0 1]	[0 1 0 (0) 0 0 0 1]	[0 1 1 1 1 0 1 1]	308,686,950
[0 0 1 0 1 0 0 1]	[1 0 1 0 0 1 0 1]	[0 1 0 1 0 0 0 1]	[0 1 1 1 0 1 0 0]	63,563,940
[1 0 1 0 1 0 1 0]	[0 0 1 0 0 1 0 0]	[0 0 0 0 0 1 1 1]	[0 0 0 (0) 1 1 0 0]	514,080

Figure 6.8 Illustrative example: the population of the next generation after random mutation and their fitness, evaluated as the product of the numerical values of the variables.

The selection of pairs and crossover operation results in the new members of the next generation Figure 6.7.

The final step is the application of mutation operator. We will assume that the mutation rate is 0.02, which means that on average one randomly selected bit in every 50 bits will be switched. In this population of six strings, there are a total of 192 bits (6 × 32), and four randomly selected bits will be switched. The bits that have been switched by random mutation are shown in circles in Figure 6.8.

The values of the fitness function for the new generation are shown in the right-hand column.

6.4 SELECTION METHODS

The random method for selecting the mating pairs for generating the next generation through recombination and crossover that was described earlier is also referred to as *Roulette Wheel Selection* method. There are also other methods of selection.

Rank Selection: This is used in situations when one or few members of the population dominate the roulette wheel. For example, one or two fittest members may have more than 90% of the roulette wheel. In this case, other members of the population have very little chance of being selected. In the rank selection method, each member of the population is assigned a number equal to their rank in the relative fitness of the population. The least fit member is assigned the rank of 1, the next fitter member 2 and so on.

The fittest member is assigned the rank of N, the number of members in the population. All members have a chance of being selected. However, this method causes slower convergence.

Steady-State Selection: In this selection method, the two least fit members of the population are replaced with offsprings generated by crossover between 2 randomly selected parents from a small group of the fittest members. All the other members move to the next generation.

Elitist Selection: The idea in this method is to assure that the top fittest member (or few members) survive to the next generation. First, the top fittest member, or few members, are copied and moved to the next generation. This is followed by the usual selection method to generate the rest of the population in the next generation.

6.5 SHAPE OPTIMIZATION OF A CANTILEVER BEAM

Consider the simple example of determining half the height h(x) of the cantilever beam of span L with rectangular cross-section, subjected to a load of P at the free end.

The height of the beam is to be determined to use the least amount of material, without exceeding the maximum allowable stress at any point, the so-called minimum weight design. This simple example happens to be a statically determinant structure (internal forces can be determined from statics alone), which means that the minimum weight design is also the fully stressed design: At every cross-section, the maximum stresses are equal to the maximum allowable stresses. This simple problem does have an analytical solution, given by the following equation, where σ_a is the allowable stress.

$$h = h_0\sqrt{1 - \frac{x}{L}}; \quad h_0 = \sqrt{\frac{3PL}{2b\sigma_a}} \quad \text{at support} \tag{6.4}$$

As can be deduced intuitively, the optimal height varies from a maximum at the fixed support to zero under the load. This is due to the fact the bending moment linearly varies between the maximum at the fixed support and zero at the free end. We will solve this problem by genetic algorithm for demonstration purposes. Note that this problem does have a single optimal solution and the problem is formulated in a finite-dimensional solution space (Figure 6.9).

Rather than solving the exact problem, we formulate a discrete version of the problem by dividing the span L into 10 equal segments and determining the height of the beam for the 10 segments. In this way, we are seeking the near optimal solution in the 10-dimensional vector space. The values of the height of the beam at the 10 points along the span are the encoded in the genetic string. For this problem, we will use a population of thirty beams.

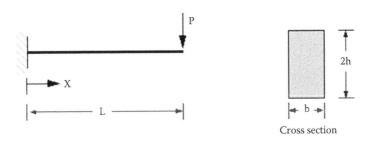

Figure 6.9 Example of the shape optimization of the tip-loaded cantilever beam.

The variation of the height of each beam along its length is encoded in its genetic string. The genetic strings are initially selected randomly.

The table labeled generation number one represents the initial randomly selected genetic codes of the population. Each line is the genetic code of one member of the population. The 10 entrees on each line are the height of the beam for the 10 segments from the fixed support (the first entry on the left) to the free end (the last entry). We have used integer values between 0 and 100. The last column lists the initial fitnesses according to the following equation:

$$f(S) = \frac{1}{\left[\left(\sum |\sigma_{max} - \sigma_a|_j\right) V\right]^2} \tag{6.5}$$

The exact form of the fitness function is not important. The basic requirement is that its value increases as we approach the optimum solution. Equation 6.5 satisfies this condition. The value of the function increases as maximum stress σ_{max} approaches the maximum allowable stress σ_a. The fitness function also increases as the volume of the material, V, decreases.

The numerical values of the fitness function shown in the table are not the actual values of the fitness function. They have been scaled such that the sum of the fitness functions for all the 30 members of the population add up to 30.

Since the genetic codes of the initial population have been selected randomly, a wide variation in the fitness functions can be observed. Some members of the population with very low value of fitness function are obviously not very good cantilever beams, since they have a thin section that causes a very high value of stress. There are four members with the initial fitness values of greater than one, two of which are very good examples, with fitness values of greater than 10.

By using the procedure described earlier, the members of the next generation are generated from this initial population. First two members are selected randomly, weighted by the fitness function. A random crossover point is determined and the two selected members are crossed over at that point. This procedure produces the two members of the next generation. The process is repeated 15 times to generate all the members of the second generation. A typical crossover is illustrated in Figure 6.10.

The two randomly selected members of the first generation

96	71	56	22	77	23	56	72	19	74
72	53	67	62	36	55	45	54	81	59

Crossover point

The two new members of the second generation after after crossover

96	71	56	22	77	23	45	54	81	59
72	53	67	62	36	55	56	72	19	74

Figure 6.10 One-point crossover of a pair of integer encoded strings.

This whole process is repeated to generate the members of the next generation from the members of current generation. In the following tables, the members of few selected generations are shown.

Although randomness is present in every step of the whole process, the positive traits of the fitter members are passed on to the future generations. Consider the move from the first to the fifth generation. The heights at the support for the four members in the first generation with fitnesses greater than one are 62, 72, 75, and 60. Observe that in only five generations the heights at the support for all the 30 members of the population are only those four values. Also, the heights at the free end are either 18 or 59, which were the value for the two members in the first generation with fitness functions greater than 10.

The fitness values in the fifth generation are much closer than they were in the first generation. Now the values at the support are either 72 or 75 and at the free end they are all 18. There are still some differences in their fitnesses.

By the fiftieth generation, all the members of the population are practically the same and all the fitnesses are very close to one. When the population is so uniform, the fitness weighted selection and crossover are no longer effective in improving the fitness. From here on, most of the improvements occur through mutation. In this example, we have used a mutation rate of 0.01. Since there are a total of 300 numbers in each generation, three numbers are selected randomly and their values are replaced with random numbers. If these changes result in fitter individuals, then they pass their genetic codes to the members of the next generation. However, if these changes result in less fit individuals, they are quickly selected out in the next few generations.

The mutation does result in improvements in the fitness. However, it acts slowly. We can see that the members in the generation number 300 are all identical and their fitness values are all exactly equal to one. From generation 300 to generation 1000 some slight improvement has occurred and two numbers have changed slightly. This trend will no doubt continue in the future generation.

The profile of the beam (half the height as a function of distance from the support) for the fittest member at some selected generations is shown in Figure 6.11. Also shown in

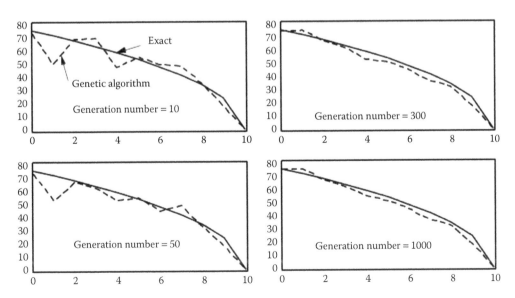

Figure 6.11 Half height of the cantilever beam at four generations in the genetic algorithm.

solid lines is the exact solution. It is evident that with the advancing generations the fittest members are approaching the exact solution. However, it may never reach the exact solution. This is what we meant in earlier discussions by the near optimal solutions and the fault tolerant nature of the soft computing methods.

GENERATION NUMBER...... = 1

Genetic codes										Fitness
1	34	39	98	85	68	11	57	1	47	0.0000
96	71	56	22	77	23	56	72	19	74	0.7614
64	69	81	14	94	60	82	77	49	70	0.1697
94	45	94	67	10	27	5	37	2	7	0.0026
62	63	50	21	75	32	32	43	68	26	1.7864
81	85	58	92	98	8	70	37	87	89	0.0321
96	47	48	16	4	35	47	22	8	98	0.0038
5	83	78	74	11	46	68	25	86	23	0.0022
18	95	60	7	49	59	29	10	18	82	0.0213
65	58	3	77	77	92	47	60	6	99	0.0003
30	60	47	73	25	59	8	90	12	35	0.1419
89	28	35	2	24	97	99	39	25	82	0.0001
75	29	34	32	97	18	60	7	14	10	0.2250
10	96	52	93	2	2	26	46	54	33	0.0001
72	53	67	62	36	55	45	54	81	59	10.9959
10	94	11	87	2	99	43	90	87	94	0.0001
17	91	28	14	45	52	89	48	64	94	0.0821
29	4	29	31	83	50	50	65	85	49	0.0012
14	50	37	7	17	22	5	54	23	43	0.0186
35	72	28	18	83	1	66	17	61	33	0.0000
75	47	82	95	53	37	88	99	81	38	4.7090
43	36	45	10	25	96	43	71	51	60	0.0867
66	4	21	72	36	84	60	11	87	69	0.0011
33	47	2	62	61	17	9	21	11	17	0.0003
3	34	68	34	31	8	80	86	10	3	0.0006
22	91	84	8	44	85	38	59	28	20	0.0302
67	76	28	12	34	34	92	64	87	66	0.1174
13	94	41	83	30	76	12	17	94	95	0.0560
52	15	59	39	48	20	35	43	75	80	0.2223
60	45	56	42	39	61	64	49	32	18	10.5316

GENERATION NUMBER...... = 5

Genetic codes										Fitness
72	45	56	42	39	61	64	49	32	18	0.7416
72	45	67	42	39	61	64	49	32	18	0.7063
72	53	67	62	39	61	64	49	32	18	1.7650
75	47	82	95	39	61	64	49	32	18	0.8721
72	53	56	42	39	61	64	49	32	18	0.9617
60	53	67	62	36	55	45	49	32	18	1.4091
60	53	67	62	39	61	64	49	32	18	1.3226
72	53	82	95	53	55	64	49	32	18	2.5002
72	45	56	42	39	61	64	49	32	18	0.7416
60	45	67	42	39	61	64	54	81	59	0.3261
72	53	67	62	36	55	45	54	81	59	0.7984
72	45	56	42	39	61	64	49	32	18	0.7416
72	53	67	42	39	61	64	49	32	18	0.9663
60	53	67	62	36	55	64	49	32	18	1.1384
72	53	67	42	39	61	64	49	32	18	0.9663
60	45	56	42	53	37	64	49	32	18	0.9936
72	45	56	42	39	61	64	49	32	18	0.7416
60	53	67	62	36	55	45	54	81	59	0.6523
60	45	56	62	53	37	64	49	32	18	1.1895
60	45	56	42	53	37	64	49	32	18	0.9936
60	45	56	42	39	61	64	49	32	18	0.7647
60	53	67	62	36	55	64	49	32	18	1.1384
60	45	56	42	36	55	45	49	32	18	0.9405
62	45	56	42	39	61	45	49	32	18	1.0169
60	45	56	42	39	61	64	54	81	59	0.3268
60	45	56	62	36	55	45	49	32	18	1.0600
60	53	67	42	39	61	64	49	32	18	0.8795
72	45	56	42	39	61	64	49	32	18	0.7416
72	53	67	62	36	55	64	49	32	18	1.4257
72	53	82	95	36	55	45	49	32	18	1.1779

GENERATION NUMBER...... = 10

Genetic codes										Fitness
75	53	67	62	36	61	64	49	32	18	0.4962
75	53	82	95	53	55	45	49	32	18	1.2543
72	53	67	62	39	61	64	49	32	18	0.6673
75	53	67	95	53	55	45	49	32	18	1.5930
72	45	82	95	53	55	45	49	32	18	0.6882
75	53	67	62	53	55	64	49	32	18	1.8406
75	53	67	62	53	55	45	49	32	18	2.8295
72	53	67	62	36	55	64	49	62	18	0.4143
72	45	67	95	53	55	45	49	32	18	0.8315
75	53	67	95	53	55	45	49	32	18	1.5930
72	53	67	62	36	55	45	49	32	18	0.6445
72	53	67	62	39	55	64	49	62	18	0.5250
75	53	67	95	36	55	45	49	32	18	0.4910
75	53	67	95	53	55	45	49	32	18	1.5930
75	53	67	62	36	55	45	49	32	18	0.6258
72	45	82	95	53	55	45	49	32	18	0.6882
75	53	67	62	36	55	45	49	32	18	0.6258
75	53	67	62	36	55	45	49	32	18	0.6258
72	53	67	95	53	55	45	49	32	18	1.4960
72	53	56	95	53	55	64	49	32	18	0.7921
72	53	56	95	53	55	64	49	32	18	0.7921
72	53	56	62	36	55	45	49	32	18	0.5993
72	53	67	62	36	55	45	49	32	18	0.6445
72	53	82	95	53	55	45	49	32	18	1.1695
72	53	67	95	53	55	45	49	32	18	1.4960
72	53	82	95	53	55	45	49	32	18	1.1695
72	45	67	62	36	55	45	49	32	18	0.4053
72	53	82	95	53	55	45	49	32	18	1.1695
75	53	67	95	53	55	64	49	62	18	0.6464
75	53	67	95	53	55	45	49	32	18	1.5930

GENERATION NUMBER...... = 50

Genetic codes										Fitness
72	53	67	62	53	55	45	49	32	18	0.9739
75	53	67	62	53	55	45	49	32	18	1.0391
72	53	67	62	53	55	45	49	32	18	0.9739
72	53	67	62	53	55	45	49	32	18	0.9739
72	53	67	62	53	55	45	49	32	18	0.9739
75	53	67	62	53	55	45	49	32	18	1.0391
72	53	67	62	53	55	45	49	32	18	0.9739
72	53	67	62	53	55	45	49	32	18	0.9739
75	53	67	62	53	55	45	49	32	18	1.0391
75	53	67	62	53	55	45	49	32	18	1.0391
72	53	67	62	53	55	45	49	32	18	0.9739
75	53	67	62	53	55	45	49	32	18	1.0391
72	53	67	62	53	55	45	49	32	18	0.9739
75	53	67	62	53	55	45	49	32	18	1.0391
72	53	67	62	53	55	45	49	32	18	0.9739
72	53	67	62	53	55	45	49	32	18	0.9739
75	53	67	62	53	55	45	49	32	18	1.0391
72	53	67	62	53	55	45	49	32	18	0.9739
72	53	67	62	53	55	45	49	32	18	0.9739
75	53	67	62	53	55	45	49	32	18	1.0391
75	53	67	62	53	55	45	49	32	18	1.0391
72	53	67	62	53	55	45	49	32	18	0.9739
72	53	67	62	53	55	45	49	32	18	0.9739
75	53	67	62	53	55	45	49	32	18	1.0391
72	53	67	62	53	55	45	49	32	18	0.9739
72	53	67	62	53	55	45	49	32	18	0.9739
75	53	67	62	53	55	45	49	32	18	1.0391
72	53	67	62	53	55	45	49	32	18	0.9739
75	53	67	62	53	55	45	49	32	18	1.0391
72	53	67	62	53	55	45	49	32	18	0.9739

GENERATION NUMBER...... = 100

Genetic codes										Fitness
72	53	67	62	53	55	45	49	32	18	0.9992
72	53	67	62	53	55	45	49	32	18	0.9992
75	53	67	62	53	55	45	49	32	18	1.0660
75	53	67	62	53	55	45	49	32	18	1.0660
72	53	67	62	53	55	45	49	32	18	0.9992
72	53	67	62	53	55	45	49	32	18	0.9992
72	53	67	62	53	55	45	49	32	18	0.9992
75	53	67	62	53	55	45	49	32	18	1.0660
75	53	67	62	53	55	45	49	32	18	1.0660
72	53	67	62	53	55	45	49	32	18	0.9992
72	53	67	62	53	55	45	49	32	18	0.9992
72	53	67	62	53	55	45	49	32	18	0.9992
72	53	67	62	53	55	45	49	32	18	0.9992
72	53	67	62	53	55	45	49	32	18	0.9992
72	53	67	62	53	55	45	49	32	18	0.9992
75	53	67	62	53	55	45	49	32	18	1.0660
72	53	67	62	53	55	45	49	32	18	0.9992
75	53	67	62	53	55	45	49	94	18	0.4217
72	53	67	62	53	55	45	49	32	18	0.9992
72	53	67	62	53	55	45	49	32	18	0.9992
72	53	67	62	53	55	45	49	32	18	0.9992
72	53	67	62	53	55	45	49	32	18	0.9992
75	53	67	62	53	55	45	49	32	18	1.0660
72	53	67	62	53	55	45	49	32	18	0.9992
75	53	67	62	53	55	45	49	32	18	1.0660
75	53	67	62	53	55	45	49	32	18	1.0660
72	53	67	62	53	55	45	49	32	18	0.9992
72	53	67	62	53	55	45	49	32	18	0.9992
72	53	67	62	53	55	45	49	32	18	0.9992
75	53	67	62	53	55	45	49	32	18	1.0660

GENERATION NUMBER...... = 300

Genetic codes										Fitness
75	76	67	62	53	51	45	36	32	18	1.0000
75	76	67	62	53	51	45	36	32	18	1.0000
75	76	67	62	53	51	45	36	32	18	1.0000
75	76	67	62	53	51	45	36	32	18	1.0000
75	76	67	62	53	51	45	36	32	18	1.0000
75	76	67	62	53	51	45	36	32	18	1.0000
75	76	67	62	53	51	45	36	32	18	1.0000
75	76	67	62	53	51	45	36	32	18	1.0000
75	76	67	62	53	51	45	36	32	18	1.0000
75	76	67	62	53	51	45	36	32	18	1.0000
75	76	67	62	53	51	45	36	32	18	1.0000
75	76	67	62	53	51	45	36	32	18	1.0000
75	76	67	62	53	51	45	36	32	18	1.0000
75	76	67	62	53	51	45	36	32	18	1.0000
75	76	67	62	53	51	45	36	32	18	1.0000
75	76	67	62	53	51	45	36	32	18	1.0000
75	76	67	62	53	51	45	36	32	18	1.0000
75	76	67	62	53	51	45	36	32	18	1.0000
75	76	67	62	53	51	45	36	32	18	1.0000
75	76	67	62	53	51	45	36	32	18	1.0000
75	76	67	62	53	51	45	36	32	18	1.0000
75	76	67	62	53	51	45	36	32	18	1.0000
75	76	67	62	53	51	45	36	32	18	1.0000
75	76	67	62	53	51	45	36	32	18	1.0000
75	76	67	62	53	51	45	36	32	18	1.0000
75	76	67	62	53	51	45	36	32	18	1.0000
75	76	67	62	53	51	45	36	32	18	1.0000
75	76	67	62	53	51	45	36	32	18	1.0000
75	76	67	62	53	51	45	36	32	18	1.0000
75	76	67	62	53	51	45	36	32	18	1.0000

GENERATION NUMBER...... = 1000

Genetic codes										Fitness
76	76	67	62	55	51	45	36	32	18	1.0000
76	76	67	62	55	51	45	36	32	18	1.0000
76	76	67	62	55	51	45	36	32	18	1.0000
76	76	67	62	55	51	45	36	32	18	1.0000
76	76	67	62	55	51	45	36	32	18	1.0000
76	76	67	62	55	51	45	36	32	18	1.0000
76	76	67	62	55	51	45	36	32	18	1.0000
76	76	67	62	55	51	45	36	32	18	1.0000
76	76	67	62	55	51	45	36	32	18	1.0000
76	76	67	62	55	51	45	36	32	18	1.0000
76	76	67	62	55	51	45	36	32	18	1.0000
76	76	67	62	55	51	45	36	32	18	1.0000
76	76	67	62	55	51	45	36	32	18	1.0000
76	76	67	62	55	51	45	36	32	18	1.0000
76	76	67	62	55	51	45	36	32	18	1.0000
76	76	67	62	55	51	45	36	32	18	1.0000
76	76	67	62	55	51	45	36	32	18	1.0000
76	76	67	62	55	51	45	36	32	18	1.0000
76	76	67	62	55	51	45	36	32	18	1.0000
76	76	67	62	55	51	45	36	32	18	1.0000
76	76	67	62	55	51	45	36	32	18	1.0000
76	76	67	62	55	51	45	36	32	18	1.0000
76	76	67	62	55	51	45	36	32	18	1.0000
76	76	67	62	55	51	45	36	32	18	1.0000
76	76	67	62	55	51	45	36	32	18	1.0000
76	76	67	62	55	51	45	36	32	18	1.0000
76	76	67	62	55	51	45	36	32	18	1.0000
76	76	67	62	55	51	45	36	32	18	1.0000
76	76	67	62	55	51	45	36	32	18	1.0000
76	76	67	62	55	51	45	36	32		

6.6 DYNAMIC NEIGHBORHOOD METHOD FOR MULTIMODAL PROBLEMS

When applied to optimization problems, SGA described earlier can only find a single optimum solution. Many optimization problems have more than one optimum; they are referred to as *multimodal optimization problems*. A number of methods have been proposed for solving the multimodal optimization problems with varying degrees of effectiveness. Here, we will describe one method that is capable finding multiple maxima.

A new simple and effective method to solve the multimodal optimization problems called Dynamic Neighborhood Method (DNM) is proposed in Limsamphancharon (2003). DNM begins with a randomly generated initial population and goes through a number of evolutionary steps. Each evolutionary step generates a new generation. Unlike the SGA where the whole population may change from one generation to the next, in DNM only two members change in an evolutionary step. At each evolutionary step, three basic operations are performed.

1. An individual is randomly selected.
2. Its active neighborhood is formed by selecting N_b−1 closest individuals.
3. Within that neighborhood, two offsprings replace the two least fit individuals.

The only parameter to be specified in the neighborhood formation is the number of individuals in a neighborhood (N_b). In creating the neighborhood, the distance between the individuals needs to be determined. This can be done in two ways: (1) using Hamming distance in genotypic space that is the space of binary strings; (2) using Euclidian distance in phenotypic space that is the solution space or real world. Hamming distance in genotypic space is the number of binary digits that are different between the two individuals. In phenotypic real-world space, the actual Euclidian distance between the two individuals is used. These two methods of measuring the distance lead to Genotypic Dynamic Neighborhood Method (GDNM) or Phenotypic Dynamic Neighborhood Method (PDNM). A small random number is added to the apparent distance to break the ties between individuals that have the same distance, especially at early stage of the evolutionary process.

An evolutionary step takes place within the neighborhood. Two parents are selected by the ranking method. Crossover between parents produces two offspring that go through mutation and replace the two least fit individuals in the neighborhood. Next, a new neighborhood is selected and the whole process is repeated. The algorithm stops when the total number of evolutionary steps reaches an allowable maximum, or when the population reaches a steady state within a specified error tolerance. It should be noted that neighborhoods in DNM are *dynamic* in the sense that they are changing with time. Late in the run, niches are formed and they independently evolve to their corresponding optima. The performance of the DNM will be illustrated next.

6.6.1 Himmelblau problem

This is one of several illustrative examples presented in Limsamphancharon (2003).DNM is applied to the modified Himmelblau optimization problem given in the following equation.

$$F(x, y) = 200 - (x^2 + y - 11)^2 - (x + y^2 - 7)^2 \tag{6.6}$$

This function has four optima with equal values of 200 at $(x_{opt}, y_{opt}) = (3.5844, -1.8481)$, $(-2.8051, 3.1313)$, $(3.0, 2.0)$, and $(-3.7793, -3.2832)$. A population of 200 is used with each individual being represented by a gray coding string of 30 bits. The first 15 bits maps the value of x from −6.0 to 6.0 and the second 15 bits maps the value of y from −6.0 to 6.0. Two-point crossover, with probability of 1.0, and mutation rate 0.005 are used. Five runs are performed for each of the four cases, that is, GDNM with $N_b = 5$ and 10 (GNDM$_5$ and GDNM$_{10}$), and PDNM with $N_b = 5$ and 10 (PDNM$_5$ and PDNM$_{10}$). All runs are terminated at the evolutionary step number 5,000. The boundaries between niches in this problem are not very well defined. This study considers an individual to be a member of niche i if it is within the rectangular region $[x^i_{opt} \pm 0.1, y^i_{opt} \pm 0.1]$.

Figures 6.12 shows the distribution of the population at the evolutionary step number 5,000 of a typical run in each case. At the end of the runs, the average percentage of individuals in all niches is 33%, 82%, 23%, and 46% for GDNM$_5$, GDNM$_{10}$, PDNM$_5$ and PDNM$_{10}$

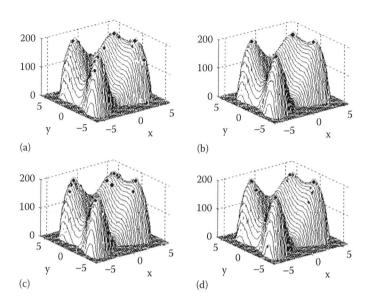

Figure 6.12 Population of DNM in Himmelblau problem at evolutionary step 5,000: (a) $GDNM_5$, (b) $GDNM_{10}$, (c) $PDNM_5$, and (d) $PDNM_{10}$. (From Limsamphancharon, N., Condition monitoring of structures by using ambient dynamic responses, PhD thesis, Department of Civil and Environmental Engineering, University of Illinois at Urbana-Champaign, Urbana, IL, 2003.)

respectively. Both $GDNM_5$ and $GDNM_{10}$ have three runs that fail to maintain one niche. Four runs of $PDNM_5$ successfully maintain all four niches but one run fails to maintain niche of the fourth optimum. All five runs of $PDNM_{10}$ are able to form and maintain all four niches. This example, as well as the other examples shown in Limsamphancharon (2003), indicate that PDNM appears to be more accurate than GDNM.

A good niching method should not only maintain the desired niches but also find the optima of every niche. Limsamphancharon (2003) presents a way of quantifying the accuracy of the solutions for multimodal functions. A parameter d_{ij} is defined as the Euclidean distance from optima p_i to the nearest individual in the population at the end of run j. For m runs of n optima problem, there will be n × m numbers of d_{ij}. The statistical measures of d_{ij}s, that is, average (d_{ave}), standard deviation (d_{std}), minimum (d_{min}), and maximum (d_{max}), can be used to quantify the accuracy of the solutions in general sense (Table 6.1).

It is clear from this table that the phenotypic DNM (GDNM) performs better than genotypic DNM (GDNM).

Table 6.1 Accuracy of DNM for the Himmelblau problem

Case	d_{min}	d_{ave}	d_{max}	d_{std}
$GDNM_5$	0.0023	0.0709	0.7140	0.1584
$GDNM_{10}$	0.0004	0.8415	6.3631	2.0419
$PDNM_5$	0.0006	0.0314	0.1662	0.0375
$PDNM_{10}$	0.0002	0.0125	0.0358	0.0121

6.6.2 Concluding remarks on DNM

A new niching method, DNM, is presented in Limsamphancharon (2003). The advantages of DNM are its simplicity and robustness. Several illustrative examples, including the one presented above, show the successes of DNM in multimodal optimization problems. DNM requires only one parameter, the neighborhood size (N_b). The results indicate that DNM is able to successfully form and maintain several niches. The results also show that PDNM performs better than GDNM in all examples presented in Limsamphancharon (2003). The superior performance of PDNM is attributed to the compatibility between definition of distance measure and definition of niches, which are both defined in phenotypic space. In contrast, GDNM attempts to maintain niches in phenotypic space by using neighborhood that is defined genotypically. When parameters of the problems are subjective, defining neighborhood in phenotypic space might not be possible, and GDNM is a good alternative.

Size of the neighborhood is an important parameter in DNM. If N_b is too large, global information can flow through the whole population, and DNM will behave like a SGA. If N_b is too small, niching capability may be increased but the evolution process may be very slow and the solutions obtained may not be accurate.

6.7 SCHEMA THEOREM

Schema theorem that was originally proposed by John Holland (Holland, 1975) is also called the fundamental theorem of genetic algorithm. Although its limitations were observed later, schema theorem does explain the power of genetic algorithm.

Genetic algorithms use a population of encoded strings rather than a single point to search for the optimal solution, but the most powerful characteristic making genetic algorithms robust and efficient in solving optimization problems is the existence of a large number of the subsets of the strings with similarities at certain string positions, called schemata. In general, for a binary string of length l and a population of size n, the total number of schemata may range from 2^i to $n.2^i$ according to the variety of the population. The above quantitative measures of schemata reveal that a tremendous amount of the information about significant similarities is really stored in a population of moderate size and is processed by genetic algorithms.

According to Holland (1975), the expected number of the schemata with better-than-average performance, short defining length, and lower order will increase exponentially in future generations after reproduction, crossover, and mutation. The conclusion is known as the Schema Theorem or the Fundamental Theorem of genetic algorithms. Further investigated by Goldberg (1989), the number of schemata processed usefully in each generation is proportional to the cube of the population size, that is, the order of n^3. Therefore, genetic algorithms intrinsically process a great number of schemata in parallel by actually handling a small amount of strings and propagate building blocks at an exponentially increasing rate generation after generation. Holland named this special observation intrinsic parallelism.

Chapter 7

Implicit redundant representation in genetic algorithm

7.1 INTRODUCTION

In genetic algorithm, the variables are assigned to the predefined segments of the genetic string. For example, in Simple Genetic Algorithm (SGA) the genetic string consists of the ordered sequence of the encoded variables and parameters of the problem. Every segment of the genetic string contains useful information. This type of representation works reasonably well for most of the simple optimization problems, which have the number of variables fixed at the outset and unchanging during the evolution. The majority of the current optimization problems belong to this class, described by the following equation:

$$\min f(x_1, \ldots, x_N) \tag{7.1}$$

subject to constraints,

$$g_j(x_1, \ldots, x_N) \geq 0; \ j = 1, \ldots, k \tag{7.2}$$

where f is the cost function, x_1 through x_N are the variables, and g_1 through g_k are the constraints of the problem.

The next level of complication arises when the number and definition of variables is dynamic. This means that some new variables may enter and some existing variable may exit the problem during the evolution. It is cumbersome to describe this kind of problem in mathematical equations. The following equations describe some form of the problem. The optimization problem

$$\min f_n(x_1, \ldots, x_N) \tag{7.3}$$

subject to k constraints,

$$g_{jn}(x_1, \ldots, x_N) \geq 0; \ j = 1, \ldots, k \tag{7.4}$$

changes to the optimization problem,

$$\min f_{n+1}(x_1, \ldots, x_M); \ M \neq N \tag{7.5}$$

subject to k constraints,

$$g_{j(n+1)}(x_1, \ldots, x_M) \geq 0; \ j = 1, \ldots, k \tag{7.6}$$

when the following condition is met:

$$h_n\left(x_1, \ldots, x_M\right) = 0 \tag{7.7}$$

As genetic algorithm is used for more and more complex problems, the vast majority of these problems will have changing number of variables. A fixed number of variables and predefinition of the segments of the genetic string may not be a viable option for most of these types of problems.

We can also imagine a situation in which a large number of solutions, perhaps infinite in number, with different numbers of variables are possible. No special conditions need to be met to move from one set of variables to another. The optimal design methodology must seek the "best" solution among all the possible solutions with different number of variables. If we artificially fix the number of variables, then the search will be confined to a subspace of the solution space, and it will reduce to a classical optimization problem. The solution space of varying number of variables can be open-ended, and a true optimal solution may exist. Unstructured engineering design problems belong to this category. The conceptual design may take many forms, each with its own set of variables. If the search is constrained to be around a given conceptual design, then one or more optimal solutions may exist. A search for a solution in an unstructured design problem normally would consist of two stages. The first stage primarily consists of design synthesis until a conceptual design is arrived at. The major part of the optimization of conceptual design takes place in the second stage.

Genetic algorithm can be an effective search method in an unstructured design solution space. It has to be formulated to allow the search to be as unrestricted as possible, without predefining the variables. One possibility is the use of *autogenesis*. We introduce this term to describe the property of spontaneous emergence of the information in the genes during the evolution. Natural genes have this property. The information content of the DNA is itself the result of the evolution. No external agent has preassigned segments of DNA to represent specific pieces of information, as we do in genetic algorithm. In fact, the emergence of life forms is itself the ultimate unstructured design problem, which is intricately coupled with autogenesis in the genes.

Implementing some form of autogenesis in genetic algorithm is likely to result in fairly powerful optimal design methods that can be applied to unstructured design problems. The key in incorporating autogenesis in genetic algorithm is in representation. How to formulate the problem for genetic algorithm without explicitly predefining the variables? There may be many different ways of achieving this goal. We will describe a method that allows some form of autogenesis.

Natural genes differ from most genetic algorithm methods in another significant way. Only a small portion of the DNA appears to contain information that is used currently. The rest of the DNA can be considered as redundant. It is possible that these redundant parts of the DNA may not be quite redundant and may contain the remnants of our past history of evolution, and as such, they may play a role in the emergence of useful information in the future. The redundant portions of the DNA may in fact play an important and crucial role in the evolution. Our experience in application of genetic algorithm in unstructured design problems has shown that redundancy in genetic strings plays a significant role in improving the performance of the method.

Unlike natural genes, most genetic algorithm methods use all the genetic string for storing useful information, and each segment of the string is assigned to a variable or a parameter of the problem. Redundant segments need to be introduced in the genetic strings before any form of autogenesis can be introduced.

Autogenesis and redundancy are interrelated, and it is difficult to think of autogenesis in genetic algorithm without redundancy. The new information in autogenesis has to emerge from the redundant segments of the genetic string. Various ways of including redundancy in the genetic code have been proposed. The simplest method is to declare and keep segments of the genetic string as redundant. Redundancy can also be dynamic. Some mechanism or operator can be incorporated in the genetic string to dynamically indicate which parts of the genetic string contain the variables and which parts are redundant. In one method, the string is divided into fixed segments and indicators are encoded at the beginning of the string to designate the segments occupied by the variables.

In Chapter 8, we will present a method used in design that allows variables in the genetic code to be turned on and off. The segments of the genetic code occupied by variables that get turned off become redundant. The redundant segments can also be turned back into active variables.

In the next section, we will present a method that allows the spontaneous dynamic emergence of the variables and parameters from the redundant portions of the string. The opposite can also occur. The segments of the string containing the variables can also dynamically turn into redundant segments.

Redundant string segment in genetic algorithm can also affect the performance of the method itself. In the fixed format genetic algorithm without any redundant capacity, the crossover operator will almost always affect some variable by breaking it up, except when the crossover point falls between two variables. This would cause the increased probability of disrupting the fit individuals. When there are redundant segments in the genetic string the crossover point may also fall in the redundant segments, in which case the variables will remain unchanged, therefore reducing the probability of disrupting the fit individuals.

7.2 AUTOGENESIS AND REDUNDANCY IN GENETIC ALGORITHM

Implicit Redundant Representation in Genetic Algorithm (IRRGA) was proposed (Raich and Ghaboussi, 1997; Raich, 1998) as a method of incorporating autogenesis and redundancy in genetic algorithm, with the intention of applying it to unstructured design problems. We will see later that IRRGA can also be used in standard optimization problems, and it can be more effective than the fixed format SGA.

In IRRGA, the number and the location of the variable of the problem are not specified at the outset; they emerge during the evolution. The individuals at any given generation in IRRGA may have different numbers of variables. The problems need to be formulated in a way that each genetic string with its own set of variables is an acceptable solution. We will see examples of this type of problem formulation.

The emergence of a variable and its location is indicated by the occurrence of a predefined sequence of bit in the genetic string. These predefined sequences of bits are called the *Gene Locator Patterns* (GLP). In binary strings, the GLPs can be a sequence of N bits of the same value, such as [1111] or [0000], or a predefined sequence of N bits, such as [0100]. Similar GLPs can be designated in other types of string encoding. For instance, in integer encoded strings GLPs can be taken as any bit values that are a multiple of 5.

In parsing the genetic strings from left to right (which is an arbitrary convention) whenever a GLP is encountered, then a predefined number of bits following the GLP is taken to represent either a single variable or a group of variables and parameters (the later will be referred to as a *gene instance* or simply a gene). In an initially random population of strings, a different number of variables will emerge in the genetic strings. In Figure 7.1, we schematically show a population of seven strings with the GLPs and variables. In this instance,

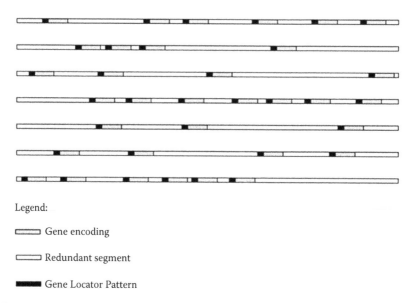

Legend:

━━━ Gene encoding

━━━ Redundant segment

▬▬▬ Gene Locator Pattern

Figure 7.1 A typical population of genetic strings in IRRGA. Gene Locator Patterns in random strings indicate the gene instances.

the number of variable ranges between 3 and 7. The average number of variables in the randomly generated initial population can be controlled by the string length.

The fitness evaluation and the genetic operators of selection, crossover, and mutation in IRRGA are the same as in the other genetic algorithm methods. The only additional consideration in IRRGA is the determination of the string length, which is directly related to the average percentage of redundancy in the strings.

7.2.1 String length and redundancy ratio

The role of redundancy in finding the maximum fitness using the IRR is influenced strongly by the total length of the string selected. The percentage of redundancy in the randomly generated initial string can be estimated by the probability of occurrence of a GLP at any position in the genetic string. We will discuss in some detail the calculation of the initial percentage of redundancy in binary-encoded strings.

The probability of an occurrence of a GLP consisting of n bits all having the same value can be determined probabilistically as defined by the following equation:

$$p = \beta^n \left[\frac{1-\beta}{1-\beta^n} \right]^2 \sum_{j=0}^{n-1} \beta^j \tag{7.8}$$

where:
   p is the probability of a single occurrence of a specific GLP,
   β is the probability of a single occurrence of the specified bit value,
   n is the number of bits specified in the GLP.

For the GLP of [1 1 1], n is 3 and β is 0.5, which is the probability of a binary bit having a value of one. The probability of the occurrence of [1 1 1] can be calculated as p = 0.0714.

The number of gene instances, each consisting of the GLP and the following encoded parameter value, initialized within a specified string length is calculated by determining the probable number of GLP that can be present within the string length and subtracting those that fall within the length of the encoded gene value, l_g. The result is the initial number of gene instances, N_0, as indicated by the following equation:

$$N_0 = \frac{(l_s - l_g + 1)p}{(1 + l_g p)} \qquad (7.9)$$

where:

p is the probability of a single occurrence of a specific GLP as defined,

N_0 is the total number of instances of GLP and encoded genes,

l_s is the number of bits in the string minus (n–1) bits to account for the end of the string,

l_g is the number of bits in an encoded gene instance.

A parameter called the redundancy ratio is defined as the ratio of the number of redundant bits and the total string length. The redundancy ratio of a binary-encoded string can be calculated by using the following equation:

$$\text{Redundancy ratio} = \frac{l_s - (l_g + n)N_0}{l_s} \qquad (7.10)$$

Consider GLP [1 1 1] and an eight-bit encoded gene ($l_g = 8$) in a binary-encoded string with the average initial redundancy ratio of approximately 0.5. In such a case, an average of 13 gene instances will be initialized in a 300-bit string; 18 gene instances in a 400-bit string; and 22 gene instances in a 500-bit string.

7.2.2 Illustrative example

A simplified example of the evolutionary computation of an IRRGA is presented to illustrate the Implicit Redundant Representation and the application of fitness evaluation, selection, crossover, and mutation. The example details one generation of a population of six binary-encoded IRR strings shown in Figure 7.2. A fundamental difference between the IRRGA and the SGA representation occurs during the decoding of the string for the fitness evaluation. The location of gene instances throughout the IRR string is defined by the GLP [1 1 1], which is shown highlighted in bold. The strings are parsed from left to right using the GLP to determine the number of gene instances to be evaluated by the fitness function along with the corresponding values of each gene instance. No overlap of gene instances is allowed. The encoded parameter value in each gene instance consists of an eight-bit binary number and is shown in a box following the GLP. The fitness value of each gene instance is obtained with the lowest binary digit designated by the bit at the right of the gene.

As expected, the genes occur at random locations through the strings, and the number of the gene instances varies between the strings. The fitness of each string, shown in the right-hand column in Figure 7.2, is defined by the product of the parameter values, represented by the following fitness function:

                                                                                    Fitness, F

[0 0 1 0 **1 1 1** 0 0 1 0 0 0 1 1 1 0 1 1 0 **1 1 1** 1 0 0 1 0 1 0 1 1 0 0 1 0 1 0 1 1]   5215

[0 0 0 1 0 0 1 0 0 0 **1 1 1** 1 0 1 0 0 0 0 1 1 0 0 0 0 1 0 0 1 1 0 0 1 0 0 1 1 1 0]   161

[0 0 1 0 1 1 0 0 0 1 0 0 1 0 0 **1 1 1** 0 1 0 0 0 1 0 1 0 1 0 1 1 0 0 0 1 1 0 1 0 0]   69

[0 0 1 0 1 0 0 1 0 0 1 0 0 1 0 **1 1 1** 0 0 0 1 1 1 0 0 0 **1 1 1** 0 0 1 0 1 1 0 0 1 0]   1232

[1 0 1 0 1 0 1 0 1 0 1 0 0 1 0 1 0 1 0 1 0 0 0 1 0 **1 1 1** 0 1 0 0 1 1 0 0 0 1 1 0]   76

[1 0 **1 1 1** 0 1 0 1 1 0 1 0 0 1 0 1 0 1 0 0 1 0 0 0 0 0 0 1 0 **1 1 1** 0 1 1 0 0 1 1 0]   9180

Figure 7.2 A population of six binary-encoded strings for IRRGA, showing the Gene Locator Patterns and the gene instances.

$$F(x_i) = \prod x_i \qquad (7.11)$$

The optimal solution and the highest fitness value correspond to case with the maximum number of genes in the string and the maximum value of each gene.

Strings are selected for the next generation based on fitness-proportionate selection. The genetic operators for crossover and mutation are applied in exactly the same manner as in SGA. Strings are paired randomly and are subjected to crossover at a single random location within the string as shown in Figure 7.3.

As can be seen in Figure 7.3, the effect of the crossover on the gene instances depends on where the cross over point falls. If it falls in the redundant segments of the pair, it can either have no effect on the gene instances, or it may destroy a GLP or create a new GLP and the trailing gene, as is the case in the first pair. The newly created gene may engulf and destroy an already existing GLP, which has also occurred in the first pair in Figure 7.3. The crossover point may fall in within the gene, and change its value, which is the case in the

String pairs 1 and 6 and crossover point after bit number 16

[0 0 1 0 **1 1 1** 0 0 1 0 0 0 1 1 1 0   1 1 0 **1 1 1** 1 0 0 1 0 1 0 1 1 0 0 1 0 1 0 1 1]

[1 0 **1 1 1** 0 1 0 1 1 0 1 0 0 1 0 1   0 1 0 0 1 0 0 0 0 0 0 1 0 **1 1 1** 0 1 1 0 0 1 1 0]

String pairs 4 and 6 and crossover point after bit number 8

[0 0 1 0 1 0 0 1   0 0 1 0 0 1 0 **1 1 1** 0 0 0 1 1 1 0 0 0 **1 1 1** 0 0 1 0 1 1 0 0 1 0]

[1 0 **1 1 1** 0 1 0   1 1 0 1 0 0 1 0 1 0 1 0 0 1 0 0 0 0 0 0 1 0 **1 1 1** 0 1 1 0 0 1 1 0]

String pairs 1 and 2 and crossover point after bit number 27

[0 0 1 0 **1 1 1** 0 0 1 0 0 0 1 1 1 0 1 1 0 **1 1 1** 1 0 0 1   0 1 0 1 1 0 0 1 0 1 0 1 1]

[0 0 0 1 0 0 1 0 0 0 **1 1 1** 1 0 1 0 0 0 0 1 1 0 0 0 0 1   0 0 1 1 0 0 1 0 0 1 1 1 0]

Figure 7.3 Single point crossover at random locations within the string applied to three pairs of fitness-proportionate selected pairs.

[0 0 1 0 **1 1 1** 0 0 1 0 0 0 1 1 1 0 0 1 0 0 1 0 0 0 0 0 1 0 **1 1 1** 0 1 1 0 0 1 1 0]

[1 0 **1 1 1** 0 1 0 1 1 0 1 0 0 1 0 **1 1 1** 0 1 1 1 1 0 0 1 0 1 0 1 1 0 0 1 0 1 0 1 1]

[0 0 1 0 1 0 0 **1 1 1** 0 1 0 0 1 0 1 0 1 0 0 1 0 0 0 0 0 1 0 **1 1 1** 0 1 1 0 0 1 1 0]

[1 0 **1 1 1** 0 1 0 0 0 1 0 0 1 0 **1 1 1** 0 0 0 1 1 1 0 0 0 **1 1 1** 0 0 1 0 1 1 0 0 1 0]

[0 0 1 0 **1 1 1** 0 0 1 0 0 0 1 1 1 0 1 1 0 **1 1 1** 1 0 0 1 0 0 1 1 0 0 1 0 0 1 1 1 0]

[0 0 0 1 0 0 1 0 0 0 **1 1 1** 1 0 1 0 0 0 0 1 1 0 0 0 0 1 0 1 0 1 1 0 0 1 0 1 0 1 1]

Figure 7.4 The new generation of six strings in IRRGA, after being subjected to the crossover operation.

second pair. Finally, if the crossover point happens to fall within a GLP, it may destroy it, leave it unchanged, or it may destroy it and create a new GLP. The population of the new generation, after the crossover operation is shown in Figure 7.4, with the new GLPs and the genes shown in boxes.

Finally, the last operation is the application of mutation. The mutation is applied at a rate of 0.02. At this rate, five randomly selected bits are switched in a population with a total number of 240 bits. The effect of the mutation operator also depends on where it occurs. If the switched bit falls within the redundant segment, it may have no effect or it may create a new GLP, with consequences similar to the crossover operator. If it falls within the gene instance, then it will only influence its value. Finally, if the switched bit falls within the GLP, it will destroy. One consequence of this may be that a new GLP will emerge from within the destroyed gene. Application of the mutation operator is shown in Figure 7.5.

The population of the binary-encoded strings in the new generation is shown in Figure 7.6, along with their fitness function values.

Later in this chapter, we will present application of IRRGA in structural condition monitoring. In Chapter 8, we will present application of IRRGA in creative structural design. Both these problems depend on dynamic variable allocation, that is, the number and definition of variables do change during the evolution. But first we want to present a problem with a fixed number of variables to demonstrate that IRRGA also performs effectively in these types of problems.

[0 0 1 0 **1 1 1** 0 0 1 0 0 0 1 1 1 0 0 1 0 0 1 0 0 0 0 0 1 0 **1 1 1** 0 1 1 0 0 1 1 0]

[1 0 **1 1 1** 0 1 0 1 1 0 0 0 0 1 0 **1 1 1** 0 1 1 1 1 0 0 1 0 1 0 1 1 0 1 0 1 0 1 1]

[0 0 1 0 1 0 0 **1 0 1** 0 1 0 0 1 0 1 0 1 0 0 1 0 0 0 0 0 1 0 **1 1 1** 0 1 1 0 0 1 1 0]

[1 0 **1 1 1** 0 1 0 0 0 1 0 0 1 0 **1 1 1** 0 0 0 1 1 1 0 0 0 **1 1 1** 0 0 1 0 1 1 0 0 1 0]

[0 0 1 0 **1 1 1** 0 0 1 0 0 0 1 1 1 0 1 1 0 **1 1 1** 1 0 0 1 1 0 1 1 0 0 1 0 0 1 1 1 0]

[0 0 0 1 0 0 1 0 0 0 **1 1 1** 1 0 1 0 0 0 0 1 1 0 1 0 0 1 0 1 0 1 1 0 0 1 0 1 0 1 1]

Figure 7.5 Application of the random mutation at a predefined rate to a population of strings in IRRGA.

	Fitness, F
[0 0 1 0 **1 1 1** 0 0 1 0 0 0 1 1 1 0 0 1 0 0 1 0 0 0 0 0 1 0 **1 1 1** 0 1 1 0 0 1 1 0]	3570
[1 0 **1 1 1** 0 1 0 1 1 0 0 0 0 1 0 **1 1 1** 0 1 1 1 1 0 0 1 0 1 0 1 1 0 1 1 0 1 0 1 1]	10648
[0 0 1 0 1 0 0 1 0 1 0 1 0 0 1 0 1 0 1 0 0 1 0 0 0 0 0 1 0 **1 1 1** 0 1 1 0 0 1 1 0]	102
[1 0 **1 1 1** 0 1 0 0 0 1 0 0 1 0 **1 1 1** 0 0 0 1 1 1 0 0 0 **1 1 1** 0 0 1 0 1 1 0 0 1 0]	83776
[0 0 1 0 **1 1 1** 0 0 1 0 0 0 1 1 1 0 1 1 0 **1 1 1** 1 0 0 1 1 0 1 1 0 0 1 0 0 1 1 1 0]	5425
[0 0 0 1 0 0 1 0 0 0 **1 1 1** 1 0 1 0 0 0 0 1 1 0 1 0 0 1 0 1 0 1 1 0 0 1 0 1 0 1 1]	161

Figure 7.6 The population of genetic strings at the beginning of the next generation and their fitness values.

7.3 SHAPE OPTIMIZATION OF A CANTILEVER BEAM USING IRRGA

In the previous chapter, we used this example to illustrate how SGA works. The cantilever beam, shown in Figure 7.7, is designed to carry the load applied at the free end.

The uniform height of the beam is not optimum, in the sense of minimum weight. This problem is to determine the height of the cantilever as a function of the distance from the fixed support. The minimum weight design in this statically determinate structure is also the fully stressed design, the maximum stress at each section caused by the bending moment is equal to the maximum allowable stress. The exact optimum shape of the cantilever is known; it is given in Equation 7.12 and shown in Figure 7.8.

$$h(x) = \sqrt{\frac{6p}{b\sigma_0}(L-x)} \tag{7.12}$$

Where σ_0 is the allowable design stress of the material.

In the previous chapter SGA was used to determine the height of the cantilever beam at 10 equidistant points along its length. The values of the heights at those 10 points, h_1 through h_{10}, were encoded in the genetic strings. It appears that this is a standard optimization

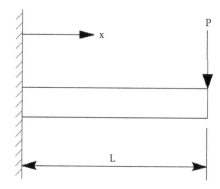

Figure 7.7 Model of cantilever beam for determining its optimum shape.

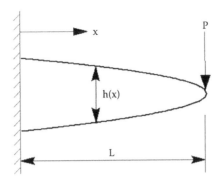

Figure 7.8 The exact optimum shape of the cantilever beam.

problem with a fixed number of variables. In this chapter, we use this example to demonstrate how IRRGA can also be applied to problems that have a fixed number of variables. Initially, this may appear to be not possible, since the number of the variables in the IRRGA cannot be specified, and they may vary from one string to the next and from one generation to the next.

In Figure 7.9, we demonstrate a situation that will arise in under specified IRRGA genetic strings. A string in this case has only three variables, x_1, x_2, and x_3, whereas the problem requires 10 variables. The determination of the 10 variables, h_1 through h_{10}, is demonstrated in Figure 7.9, where h_1 through h_4 equal x_1, h_5 through h_7 equal x_2, and h_8 through h_{10} equal x_3.

Similarly in Figure 7.10, we show a situation in which the number of variables in the IRRGA string is larger than 10. In this case, two of the variables in IRRGA string will not be used. As shown in Figure 7.10, h_1 through h_5 are equal to x_1 through x_5; x_6 is skipped; h_6 through h_{10} are equal to x_7 through x_{11}; and finally x_{12} is ignored.

This example clearly shows that the IRRGA, with autogenesis and redundancy, can also be used in standard structured optimization problems, where the number of variables is dynamically determined. If there is an advantage in a specific number of variables, then the method will move toward producing that specific number of variables. In this example, the optimum solution needs 10 or more variables. If the redundancy ration is sufficiently large, then IRRGA strings will gradually move toward having 10 or more variables.

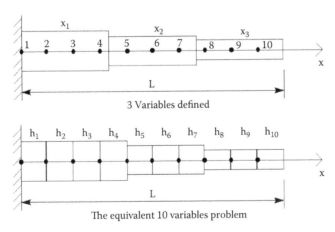

3 Variables defined

The equivalent 10 variables problem

Figure 7.9 Determining the ten variables from the underspecified three variables in the IRRGA genetic string. (From Raich, A.M. and Ghaboussi, J., *Int. J. Evol. Comput.*, 5, 277–302, 1997.)

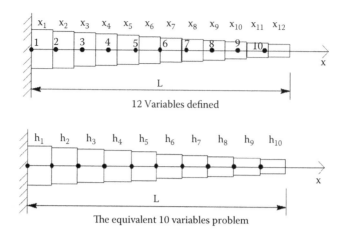

12 Variables defined

The equivalent 10 variables problem

Figure 7.10 Determining the ten variables from the overspecified 12 variables in the IRRGA genetic string. (From Raich, A.M. and Ghaboussi, J., *Int. J. Evol. Comput.*, 5, 277–302, 1997.)

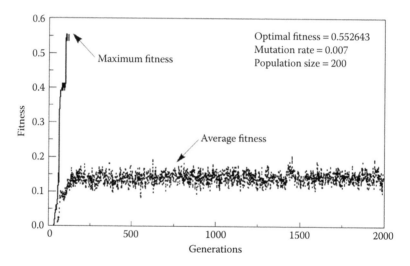

Figure 7.11 Maximum and average fitness values obtained for a binary-encoded IRRGA. (From Raich, A.M. and Ghaboussi, J., *Int. J. Evol. Comput.*, 5, 277–302, 1997.)

The application of IRRGA to the cantilever problem is discussed in detail in a paper (Raich and Ghaboussi, 1997), and also in the doctoral dissertation of Dr. Raich (Raich, 1998). Here we will present some of the results. Figure 7.11 shows the maximum and average fitness values for the binary-encoded IRRGA applied to the cantilever problems. Shown in this figure are the results of best of 10 runs. Since the initial genetic strings are generated randomly, the performance of the method will vary from one run to another, and it is customary to show the best of a number of runs. A population of 200 strings, with string lengths of 500, is used for 2000 generations. The maximum fitness is reached in 103 generations.

The same problem was also solved with SGA with the population size of 200 continued up to 2000 generations. Because of the fixed variable format, the string lengths were 80 bits to encode the 10 variables, each encoded with 8 binary bits. In the best of the 10 runs, the maximum fitness and the optimal solution was reached in 255 generations.

Comparison of the performance of IRRGA and SGA on this simple example shows two interesting significant points. The IRRGA converges much faster than the SGA. The second point is that the average fitness in the IRRGA remains lower than in SGA. This is an indication that IRRGA is able to maintain a higher level of genetic diversity than SGA. The higher level of genetic diversity means that newer forms can emerge more easily and a more efficient search of the solution space can take place in search of the optimum solution.

The performance of IRRGA is obviously influenced by the string length, redundancy ratio, and the population size. The string length and the redundancy ratio are closely inter-related. Figure 7.12 shows the value of the redundancy ratio, which earlier was defined as the number of noncoded (redundant) bits in a string divided by the total number of bits in a string. For the larger string lengths of 500 and 600 bits, the redundancy ratio increases and reaches a more or less stable value. In the case of the shorter string length of 110 bits, the redundancy ratio tends to decrease as the method seeks to find more variables. The appropriate string length and the redundancy ratio are strongly problem dependent.

The effect of the string length on the performance of the IRRGA is shown in Figure 7.13, which shows the average of the maximum fitness of 10 runs starting from random initial conditions. The string lengths were varied from 110 to 600. The longest string lengths of 500 and 600 bits give the best performance with very similar results. The performance tends to deteriorate as the strings are made shorter. In the case of the string length of 110 bits, obviously there is not enough space for the emergence of an adequate number of variables and the performance is very poor.

The effect of the population size on the performance of IRRGA is demonstrated in Figure 7.14. The situation is not very different than the other genetic algorithm methods. Since the genetic algorithm has a strong element of randomness in its search for the optimal solution, the larger population sizes will have a better chance of finding the optimum solution faster. Again, the results are presented in terms of average of the maximum fitness obtained from the 10 random trials. The population size of 2,000 arrives at the maximum fitness in 121 generations that is a relatively small number of generations. On the other hand, the fitness in the case of the population size of 100 seems to reach a plateau well below the maximum.

There are trade-offs in relation to the optimal population size for a given problem. The larger population sizes tend to converge faster, requiring smaller number of generation.

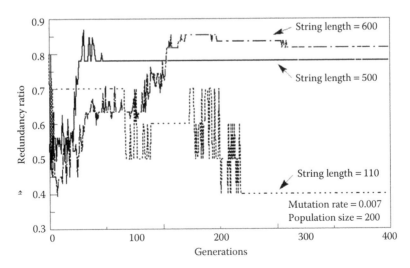

Figure 7.12 Variation of the redundancy ratio of the fittest member in the population of binary IRRGA for the cantilever beam optimization problem. (From Raich, A.M. and Ghaboussi, J., *Int. J. Evol. Comput.*, 5, 277–302, 1997.)

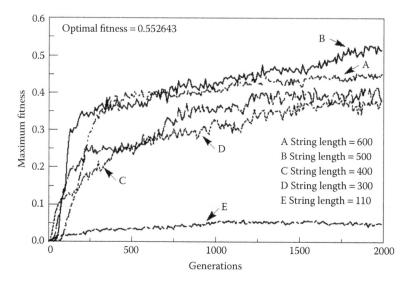

Figure 7.13 Effect of variation of string length on performance of binary-encoded IRRGA on the cantilever beam optimization problem. (From Raich, A.M. and Ghaboussi, J., *Int. J. Evol. Comput.*, 5, 277–302, 1997.)

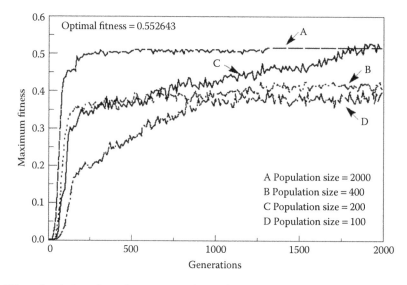

Figure 7.14 Effect of variation of population size on the performance of binary-encoded IRRGA on the cantilever beam problem. (From Raich, A.M. and Ghaboussi, J., *Int. J. Evol. Comput.*, 5, 277–302, 1997.)

On the other hand, the larger population size also requires a larger number of fitness evaluations that can be quite costly, in terms of computational resources.

7.4 IRRGA IN NONDESTRUCTIVE EVALUATION AND CONDITION MONITORING

Evaluating the condition of structures is to ascertain their safety and serviceability. No matter what causes the defect and damage in structures, naturally or man-made, it degrades the serviceability of structures. The procedure to identify changes in the structural characteristics plays a key role in detecting and assessing the state of damage and/or deterioration. To repair and retrofit, a qualitative and/or quantitative evaluation procedure has to be established to accomplish this task.

It is well known that one way of detecting the flaws in a solid body is by tapping it and listening to the emitted sound. It is an intuitive way of nondestructively evaluating the condition of a solid body. The condition of a structure, which is more complicated than a solid body, can also be evaluated nondestructively.

Testing and analysis are two major parts in Nondestructive Evaluation (NDE). For testing, several NDE techniques, such as visual inspections, dye-penetrant methods, radiographic techniques, ultrasonic tests, acoustic emission methods, and laser tests, have been successfully applied to evaluate the structural condition. But analyzing and interpreting the test result is the most challenging part of NDE. In fact, NDE problems are inverse problems in civil engineering. Inverse problems were discussed in Chapter 4. We can see that NDE is a type-two inverse problem and soft computing methods are highly effective in solving these problems. In this section, we will discuss the application of IRRGA in condition monitoring of structures.

7.4.1 Condition monitoring of a truss bridge

The objective of structural condition monitoring is to construct the qualitative and/or quantitative description of the changes in a physical structural system from the measured loads and the measured responses of the structure to those loads. There are three main issues in the condition monitoring: the existence, location, and extent of the changes of structural conditions. In general, it is more difficult to find the location and extent of the damage than its existence. System identification techniques can be used to perform this task. The parameters in the system model can be used to represent the condition quantitatively. For example, the characteristic properties of structural members, such as Young's moduli and cross-sectional areas, can be used to represent the structural condition. The structural properties can be identified by minimizing the difference between the measured response and the response computed from the computational model of the structural system. The variation in structural conditions can be detected by comparing the properties between successive monitoring episodes through the measured response.

Some studies of the damage detection use measured dynamic responses. We present examples that use static displacements as measured responses. One of the advantages of using static response in the damage detection is that static responses are more locally sensitive than natural frequencies in structural damage detection.

The method described is general and can be applied to different types of structures. We will present an example of its application, the skeletal structures, and more specifically, a truss bridges. In the actual implementation of this method, the bridges will be instrumented and remotely sensed. Passing of vehicles over bridge can be treated as a series of static load tests. The remotely sensed displacement measurements under the normal traffic can be used to identify the state of bridges, including detection of any damage. In this way, the condition of bridges can be periodically monitored under normal traffic loads.

It is assumed that only the geometry of the structure is known and its member properties including, elastic moduli, cross-sectional areas, and moments of inertia, are to be determined from the condition monitoring. The computational model of the structure is developed from the information on the geometry of the structure. The member properties in the computational model of the structure are treated as unknowns to be determined from the condition monitoring.

In the periodic determination of member properties of the structure, sudden changes in the member properties from one monitoring episode to the next are indications of the

damage in the members. More gradual changes in the member properties between successive monitoring episodes would indicate deterioration and wear. Practical considerations
normally limit the number of the displacements that can be measured.

The example presented here is from Chou (2000) and Chou and Ghaboussi (2001). The truss
bridge and the truck load are shown in Figure 7.15. It is 100-ft span and has 26 elements. It is
modeled as a plane truss. The loading used in the condition monitoring is by a loaded truck passing over the bridge at equal interval of moving from one sensor to next as shown in Figure 7.16.

Further details are given in Chou (2000) and Chou and Ghaboussi (2001). In these references, three cases with one, two, and three members being damaged have been studied.
The cases of determining the member properties with SGA and IRRGA were studied, and

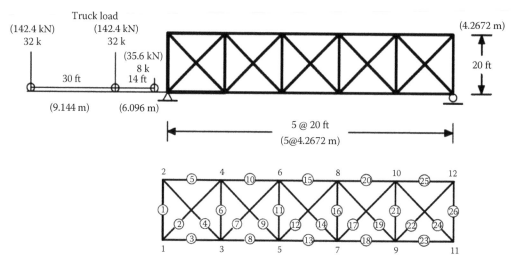

Figure 7.15 The truss bridge and the truck load passing over the bridge used in the example on condition
monitoring under normal traffic, and finite-element model of the bridge. (From Chou, J.H. and
Ghaboussi, J., *Int. J. Comput. Struct.*, 79, 1335–1353, 2001.)

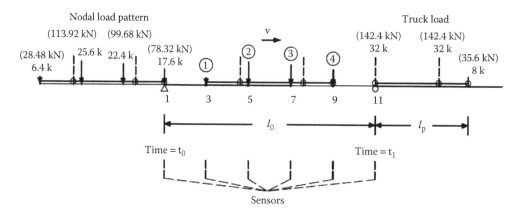

Figure 7.16 Loading conditions of truck passing over the bridge illustrating the determination of loads at
the structural node from the position of the truck. (From Chou, J.H., Study of condition monitoring of bridges using genetic algorithms, PhD Thesis, Department of Civil and Environmental
Engineering, University of Illinois at Urbana-Champaign, Urbana, IL, 2000.)

it was observed that IRRGA performs far better than SGA. In all the cases studied, it was assumed that only few displacements are measured, as that is most practical. It was also reported studying the following cases.

1. It was assumed that loading is known, as is the case in controlled tests. In addition to member properties, the unmeasured displacements were also encoded in the GA strings and determined.
2. In condition monitoring under normal traffic loads, the loads are not known. In addition to member properties, unknown loads were also encoded in the GA strings and determined.
3. All the cases described above were also studied in the presence of different level of noise in the measured displacements.

Here, we will describe only one case. It is assumed that members No. 1 and 11 are damaged and their axial stiffness is reduced by 50%, and all the other members are undamaged. It is also assumed that few displacements are measured by remotely sensing under normal traffic. So, the loads are not known. There is also noise in the measured displacements.

Genetic strings are a combination of IRRGA followed by SGA, as shown in Figure 7.17. The IRRGA is used to encode the member properties. This is followed by a segment of SGA that encodes the unknown truck loads. As shown in Figure 7.15, the truck load consists of three forces; it is assumed that the distances between these forces are known. The values of the three forces are unknown, and it depends on if the truck is loaded or unloaded and the type of load.

In the IRR part, the string lengths are determined so that the average redundancy ratio is 0.6. This will assure that on the average sufficient number of member properties will emerge. The members that are not present in any IRRGA string are assigned the baseline properties. It is also possible that more than one set of properties may emerge for a member. In that case the average of the emerged properties is used.

In these studies, simulated measurements determined from the finite-element simulations are used. Ten runs of up to 200 generations are performed to enable the statistical analysis of the results. Figure 7.18 shows the results. Figure 7.18a shows the damage index that is the ratio of the computed and the actual member properties. Figure 7.18b shows the computed truck loads. The mean value from 10 runs seems to determine the damage indices for members 1 and 11 reasonably well. There seems to be more variation in the damage index of member 11 than in member 1. We also observe a large amount of scatter in the damage indices of the undamaged members. The truck loads are determined with reasonable accuracy.

The source of the scatter in the damage indices of the undamaged members is the 10% noise in the measured displacements. There are a number of methods for removing the noise from the results of the condition monitoring. In these studies, a threshold was established.

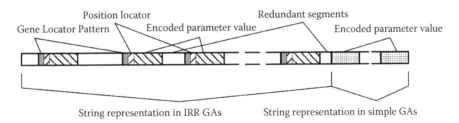

Figure 7.17 Mixed string representation. (From Chou, J.H., Study of condition monitoring of bridges using genetic algorithms, PhD thesis, Department of Civil and Environmental Engineering, University of Illinois at Urbana-Champaign, Urbana, IL, 2000.)

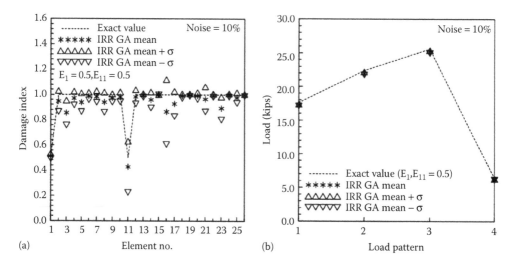

(a)   Element no.   (b)   Load pattern

Figure 7.18 IRRGA in condition monitoring of the truss bridge with members I and II damaged 50% and with 10% noise in the measured displacements: (a) damage indices from ten analyses, (b) truck loads from 10 analyses. (From Chou, J.H. and Ghaboussi, J., *Int. J. Comput. Struct.*, 79, 1335–1353, 2001.)

Only members with damage indices exceeding the threshold are considered as damaged. From the 10 analyses *probability of detecting the damage* was also determined. Damaged members should also have probability of detecting the damage higher than 50%. These two criteria are used to decide on which members were damaged.

The threshold in this study is based on 10 runs on the undamaged structure. The average of the minimum of the damage indices is taken as the threshold. In the damaged structure, only those members with the average damage index below the threshold are considered as damaged. The damage indices of the undamaged structure are shown in Figure 7.19.

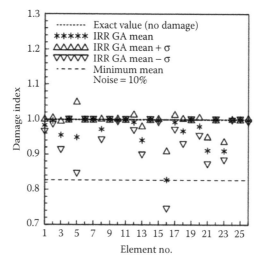

Element no.

Figure 7.19 Damaged indices for the undamaged structure. (From Chou, J.H. and Ghaboussi, J., *Int. J. Comput. Struct.*, 79, 1335–1353, 2001.)

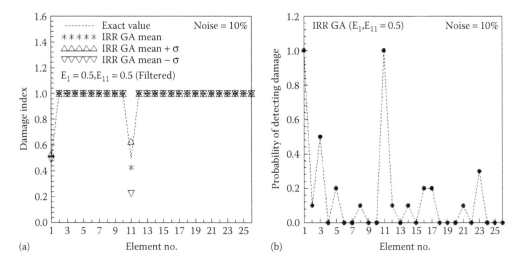

Figure 7.20 The filtered results for the plane truss bridge with 50% damage in elements 1 and 11 with 10% measurement noise in the measured displacements: (a) filtered damage index and (b) probability of damage detection at the given threshold. (From Chou, J.H. and Ghaboussi, J., *Int. J. Comput. Struct.*, 79, 1335–1353, 2001.)

With this threshold, it is also possible to determine the probability of detecting the damage from the 10 analyses. The damage indices are filtered by using the two criteria: the threshold and exceeding 50% of the probability of detecting the damage. The filtered results are shown in Figure 7.20.

The mean value of the damage index in member 1 is very close to 50%, whereas the damage index for member 11 exceeds 50%. Probability of detecting damage is 100% for members 1 and 11. Probability of detecting damage for most of the other members is below 50%, except for member No. 3. However, the damage index for this member does not exceed 50%, so it does not satisfy both criteria to be considered damaged.

Chapter 8

Inverse problem of engineering design

8.1 INTRODUCTION

Engineering design is the task of using the scientific and engineering knowledge to create new objects and structures to meet prescribed serviceability, safety, and esthetic requirements in meeting specific needs. In the current chapter, we will explore the potential application of the soft computing methods in engineering design. We are especially interested in the potential application of the evolutionary-based methods, such as genetic algorithm. Since genetic algorithm attempts to mimic the natural evolutionary process that has resulted in the "design" of the higher level live, it is natural to think that some advanced versions of genetic algorithm will have the capability of evolving solutions for engineering design problems. Although such a potential exists, we are far from having genetic algorithm-based methods for engineering design. Recent applications of genetic algorithm are concentrated more on optimization than design, and most researchers increasingly consider the genetic algorithm as an effective tool for optimization problems. We will explore the major differences between optimization and engineering design.

Engineering design is far more complex than the optimization problems. Of course, optimization can be a part of the process of engineering design, and it is a primary requirement that the engineering designs be optimum is some sense. Engineering designs are intended to be as economical as possible to manufacture, construct, and maintain, while satisfying the basic safety and the serviceability requirements. However, the role of optimization in engineering is complex. The classical mathematically based methods on their own cannot be easily applied to engineering design problems. Genetic algorithm is an exception. Although it is often used as an optimization method, it has the potential of being successfully applied to engineering design, as we will see later in this chapter.

In considering the potential application of genetic algorithm in evolving engineering designs, it is important to first understand the process of engineering design in practice. Engineering design is primarily based on experience, and it goes through phases. In contrast, a method like genetic algorithm does not use experience. It relies on random search and competition. A better understanding of the experience-based engineering design is likely to help one clarify how designs evolve in genetic algorithm.

The practice of engineering design is highly dependent on experience. Engineering designers have access to the accumulated past experience in their profession and the lessons learned from the study of past failures and successes. This is often supplemented by the designers own accumulated past experience of working on similar designs. The design experience is often used to synthesize a conceptual and preliminary design that is further refined to produce a final design.

The experience-based design can be roughly divided into conceptual design and final design. This may be an overly simplistic view. The actual design process often goes through many stages and cycles through preliminary design, evaluation, and redesign. However, it is useful to separate the preliminary and conceptual stages of the design from the task involved in the finalizing of the design.

An experienced engineer on encountering a design problem will automatically conceptualize some candidate designs on the basis of his or her past experience in dealing with similar problems. He/she would then proceed to evaluate those designs and narrow the choices by elimination. Further refinements and eliminations would lead to the choice for the final design. Finalizing the design involves the selection of the values for the parameters and variables of the design. This step often includes some form of optimization. It may be an ad hoc experience-based optimization or an application of some mathematically based optimization method.

As an example, we can consider the design of the steel frame for a high-rise building. During the conceptual and preliminary design, decisions are made about the type of framing to be used for carrying the vertical loads, such as the dead loads and live loads, and for resisting the lateral loads, such as wind and earthquake loads. The choices may include various types of moment resisting steel frames with different types of lateral bracing, or various types of tubular steel frames. After the steel frame type has been selected and its configuration and topology has been fixed, the final stage of the design involves the selection of the member types and their cross-sectional properties. In the first part of the design process, the conceptual design stage, a new frame structure is created. As a result, there is considerable room for innovation and creativity. The variable set for the problem emerges during the conceptual design stage, and it is completely defined at the end of this stage when topology and the configuration of the design are finalized. On the other hand, the final stages of the design process, the selection of the member section properties, are a simple optimization problem with a fixed number of variables.

Accumulation of past experience in engineering design combined with the gradual improvements and incremental innovations over time actually amounts to a form of evolution. Examples of this type of "industrial evolution" are gradual improvements over the past decades in products such as automobiles and in services such as telephone. A form of "industrial selection" pressure, similar to natural selection, is at work here. A primary requirement for this industrial selection is the competition between alternative designs offered to the public, and the profit motive forces the competing manufacturers to offer new and improved products that better satisfy the public's needs. Improvements seldom take place in monopolies in absence of any competitive pressure to improve. The primary difference between this type of industrial selection and the natural selection is that the former lacks the random search provision of the latter. As such, the industrial selection only searches a limited region in the solution space. A genetic algorithm-based method that has the capability of performing a random search of much larger region of the solution space has the potential of evolving innovative designs.

Innovation does occur in engineering design. However, it is often incremental and in some cases their accumulation over time may lead to major technological change. Sudden major technological innovations are rare. In large government projects, such as major new weapons systems or the space program, the designers have to move beyond the experience base, into new and unexplored territory. Such major projects often lead to major technological innovations in military as well as in civilian sectors.

Truly creative designs break new ground in creating engineering designs that are fundamentally different than any designs in existence. Since a creative design by definition has new features that do not exist in the experience base, it cannot be the result of a purely experience-based design. Creative designs use the past experience to go beyond what is contained in the past experience base to break new ground and create new designs. Creativity in structural design is more commonly attributed to the field of architecture than structural engineering. However, although truly creative and innovative designs are rare in engineering, they do occur. There are many examples of creative and innovative bridge designs. Similarly, there have been many innovations in tall building design. A good example of creative design in tall buildings is the development of tubular steel frame structures.

Creative designs can be thought of as being the result of a search in solution space that is larger than the solution space of cases contained in the past experience base. The past experience case can be thought of as the subset of the solution space where the search for creative design is taking place.

An interesting question is whether the evolutionary-based methods such as genetic algorithm have the capability of evolving truly creative designs. It is hard to think that the simple genetic algorithm can be used to evolve creative engineering designs. However, evolutionary-based methods do perform random searches of the solution space. As such, the evolutionary-based methods have the potential of evolving creative designs, if they are formulated to perform a search in an open-ended solution space not constrained by the past experience.

When using the genetic algorithm like methods in design, ideally the objective could be to evolve a new design as independent of our experience base as possible. This is especially true if we expect the methodology to evolve new and innovative designs. The evolutionary-based methodology must be able to search the solution space in its entirety. Our experience base only covers a small subspace of all the possible solutions, and any biases towards our experience base would restrict the search for innovative solutions. At the other extreme, when most of the design decisions have been made on the basis of the designer's experience, the role of the genetic algorithm approaches an optimization problem where the search occurs in small subspace of the space of all the possible solutions.

In this chapter, we will consider some forms of genetic algorithm that have been especially developed to handle design problems with minimal influence from our experience base. For engineering design problems, the genetic algorithm must move beyond the fixed format representation and static variable allocation to the more flexible *dynamic variable allocation*. We will describe two specific implementations of dynamic variable allocation from the research work of the author and his coworkers and present applications of these methods in two examples taken from two doctoral dissertations.

8.2 STRUCTURED AND UNSTRUCTURED DESIGN PROBLEMS

The *unstructured designed problem* is a term introduced by the author to represent the class of problems that lie somewhere between a simple optimization problems and the wide-open design problems. As we have seen earlier, the simple optimization problems are those with a fixed number of variables, and the optimum solution is sought in a finite-dimensional solution space. On the other hand, the variable definition in a design problem is dynamic and the solution space is open-ended, and it is potentially an infinite-dimensional solution space. As the designers make choices during the design process, the solution space is reduced to smaller and smaller subspace of the open-ended solution space.

Unstructured design problems arise at an intermediate stage between a completely open-ended design problem and a simple optimization problem. Some design decisions have been made. However, there is considerable flexibility in the problem definition to allow some creativity. In fact, the term unstructured design problem was coined in the course of the

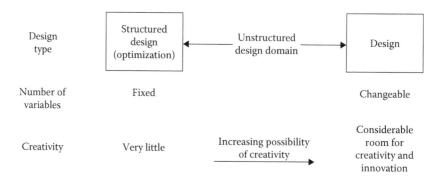

Figure 8.1 Illustration of the unstructured design problem domain in relation to the structured design (optimization) and the open-ended design.

research by the author and his associates to explore the possibility of the emergence of the creative designs in evolutionary models.

The domain of the unstructured design problems is illustrated in Figure 8.1. In the structured design problems, there is very little opportunity for innovation and creativity. A fixed number of variables are selected to perform a classical optimization. At the other extreme is the open-ended design, where the number of variables is changeable and dynamic and their definitions may also be changeable. Innovative designs are possible by creating designs well beyond the domain of the past experience.

We will illustrate the optimization, design, and the unstructured design problems by a simple example. The problem is bridging a fixed span, L. The end points of the span are given, and it is assumed that the set of loads to be supported by the bridge is also known.

In this case, a structured design problem (or an optimization problem) arises when most of the decisions about the material and structural type have been made and a fixed number of variables have been defined for the optimization problem. The task is to determine the values of those variables to minimize some cost function, subject to some constraints. An example is shown in Figure 8.2a. A truss structure has been designed and the objective of the optimization problem is to determine the cross-sectional areas of the truss members to minimize the total volume of the material.

It is not intended to imply that this is a simple task. What we are calling the simple optimization problem can indeed be a very difficult problem to solve with the classical mathematically based optimization methods. On the other hand, these problems can be solved with the fixed format Simple Genetic Algorithm (SGA), where variables are assigned to fixed portions of the genetic code.

The open-ended design problem is depicted in Figure 8.2c. The choice of the structural type and material are part of the design problem. The designer can choose from a number of materials, such as wood, reinforced concrete, steel, composite material, a combination of the above, or some new type of material. There are many structural types available from the past experience. They include steel plate girder or box girder, reinforced concrete open girders or box girders, various types of arches, various types of trusses, suspension structure, and many other bridge types. In a truly open-ended design problem, the possibilities also should include all the designs that have not been discovered yet. An evolutionary-based random research may come up with a completely new design.

The unstructured design problems fall in the domain between the straightforward optimization problems and the completely open-ended design problems. This covers a very wide range. In an unstructured design problem, some design decisions have been made. However,

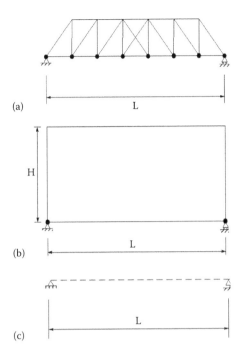

Figure 8.2 The design problem of bridging a fixed distance L to carry the fixed loads: (a) structured design problem (optimization), task: Determine the section properties to minimize the total volume of the material; (b) unstructured design problem, task: Determine the best possible safe and economical truss bridge within the physical design space; (c) open-ended design problem, task: Design the best possible safe and economical bridge.

the scope of the remaining part of the design leaves considerable room for innovation and creativity in design. Figure 8.2b depicts an unstructured design problem. The decision has been made that the span L will be bridged by a truss and that truss will be designed to lie within a predefined physical design space. However, no decision has been made about the truss type, its topology, or its configuration. The number and location of the nodes and number and location of the members are all unknown at the outset. The scope of such an unstructured design problem leaves considerable room for innovation and creativity.

A solution of the unstructured design problem shown in Figure 8.2 using genetic algorithm will be presented in a later section in this chapter. We will also present an unstructured design of a building frame structure. The location of the floors, the load cases (dead load, live load, and wind load), and the physical design space will be specified. The location, length, and inclination of the load-carrying members will not be specified.

The fixed format representation and static variable allocation genetic algorithm are not suitable for the unstructured design problems. The number of variables and their definition are not known at the outset; they will evolve during the evolution. In the case of the unstructured design of the truss bridge, the number of nodes and members of truss can change. It is likely that different members of population at any given generation will have a different number of nodes and truss members. The unstructured design problems require *dynamic variable allocation* in the genetic algorithm. The representation of the problem in genetic algorithm must allow the spontaneous appearance and disappearance of nodes and truss members during the evolution. The implementation of dynamic variable allocation in genetic algorithm is the subject of the next section.

8.3 DYNAMIC VARIABLE ALLOCATION AND REDUNDANCY IN GENETIC ALGORITHM

In most applications of SGA, the variable allocation is static; that is, the variables or substrings are permanently assigned to segments of the genetic string. The variable allocation is the same for all the members of the population in a generation and the locations of variables or substrings do not change from one generation to the next. In the simplest form of the static variable allocation genetic algorithm, the string lengths are fixed and the whole string is assigned to variables or substring, leaving no unused segments in the genetic string. In other variations of the static variable allocation genetic algorithm the strings may include fixed unused or redundant segments, or the string length may vary within each generation and from one generation to the next.

The static variable allocation is appropriate for most optimization problems where the number and definition of the variables are known at the outset. Fixed string length genetic algorithm with and without the redundant segments can work in these problems. In some cases, the identity of the variables does not change but the number of variables may increase during the evolution. Similarly, some variables may disappear. Variable string lengths can be used for these types of problems.

The static variable allocation methods are not suitable for the unstructured design problems. The number and the definition of the variables in the unstructured design problems can change during the evolution. This requires a form of dynamic variable allocation to allow the emergence of new variables and disappearance of some of the existing variables during the evolution. There are many possible methods for implementing dynamic variable allocation. We will discuss two methods that were developed and utilized by the author and his doctoral students. These two methods will be used in the examples presented in this chapter.

The first method for implementing dynamic variable allocation is a variation of the fixed length static variable allocation. The fixed length of the strings is permanently allocated to substrings. A number of bits in each substring are designated as an on/off switch and the remainder of the substring contains one or more variables. The variables in the substring may define a specific entity in the unstructured design problem, such as a node or a member in structural design problem. The on/off switches indicate the status of the substrings. A predefined rule transforms the bits in the on/off switch to a binary value indicating that the substring is either on or it is off. Only the substrings which are on are decoded to formulate the unstructured design. The substrings that are off are treated as redundant and unused segments in the genetic string. This version of the dynamic variable allocation is illustrated in Figure 8.3 for a small number of individuals in a generation. The number of active substrings may vary within a generation.

The method of dynamic variable allocation with the on/off switches is illustrated Figure 8.4 on a population of four members. The binary strings of 40 bits are divided into five substrings. The first two bits in each substring constitute the on/off switch and the next six bits encodes a single variable. The variable is active if the two bits in the on/off switch have the same value, either [1, 1] or [0, 0]. Otherwise, substring is inactive. The active substrings in Figure 8.4 are shown in bold. The first two strings each have three variables and the last two strings each have two variables.

The method used in the interpretation of the on/off switch in the illustrative example shown in Figure 8.4 would result in a redundancy ratio (the ratio of redundant bits to the total number of bits in the string) of 0.5 in the initial random population. A larger number of bits can be used in the on/off switch, and there are many different ways of interpreting those bits. In the example presented in the next section, on/off switches had 10 bits and the switch was interpreted as being on when 5 or more bits had values of 1. The appropriate number

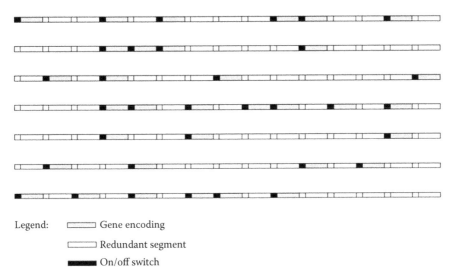

Legend: ⊏⊐ Gene encoding

⊏⊐ Redundant segment

▬▬ On/off switch

Figure 8.3 Schematic illustration of implementation of dynamic variable allocation using on/off switches.

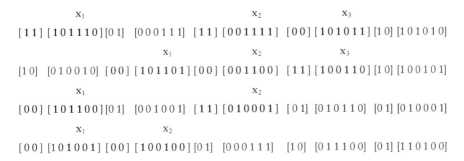

x_1 x_2 x_3

[1 1] [1 0 1 1 1 0] [0 1] [0 0 0 1 1 1] [1 1] [0 0 1 1 1 1] [0 0] [1 0 1 0 1 1] [1 0] [1 0 1 0 1 0]

x_1 x_2 x_3

[1 0] [0 1 0 0 1 0] [0 0] [1 0 1 1 0 1] [0 0] [0 0 1 1 0 0] [1 1] [1 0 0 1 1 0] [1 0] [1 0 0 1 0 1]

x_1 x_2

[0 0] [1 0 1 1 0 0] [0 1] [0 0 1 0 0 1] [1 1] [0 1 0 0 0 1] [0 1] [0 1 0 1 1 0] [0 1] [0 1 0 0 0 1]

x_1 x_2

[0 0] [1 0 1 0 0 1] [0 0] [1 0 0 1 0 0] [0 1] [0 0 0 1 1 1]   [1 0] [0 1 1 1 0 0] [0 1] [1 1 0 1 0 0]

Figure 8.4 An illustration of dynamic variable allocation with the on/off switches.

of bits in the on/off switches depends on the length of the substrings. The longer substrings generally would require more bits in the on/off switch. This relationship is not precise and the details of the method of interpretation of the on/off switch should not affect the outcome of the problem as long as a reasonable level of initial redundancy ratio is achieved.

Another example of the implementation of dynamic variable allocation is the Implicit Redundant Representation in Genetic Algorithm (IRRGA) method that was described in Chapter 7. This method was specially designed to allow the emergence of new variable and disappearance of some existing variables. The substrings are preceded by the appearance of a predefined sequence of bits (Gene Locator Patterns) in the string. The specific Gene Locator Patterns used in this book is a sequence of three ones [1 1 1]. Whenever such a sequence is encountered in parsing the string, the subsequent bits are taken as the substring. This method is illustrated schematically in Figure 8.5.

In IRRGA, the substrings can appear anywhere within the string, unlike the first method of dynamic variable allocation shown in Figure 8.3 in which the location of the substrings was fixed. The strings in IRRGA can develop a different number of substrings, as illustrated in Figure 8.5.

Both methods of dynamic variable allocation described in this section were originally developed for applying to the unstructured design problems, and they have proven to be

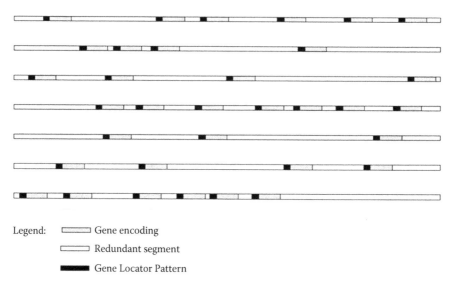

Legend: ⊏▭⊐ Gene encoding

⊏▭⊐ Redundant segment

▬▬ Gene Locator Pattern

Figure 8.5 Schematic illustration of the dynamic variable allocation in the Implicit Redundant Representation Genetic Algorithm.

effective methods for this class of problems. However, dynamic variable allocation has also proven to be effective in the standard problems with a fixed number of variables.

The role of redundancy in genetic algorithm is very interesting, and it has not been studied extensively. The author's experience shows that redundancy can play an important role in improving the effectiveness of the genetic algorithm. For example, we have shown in Chapter 7 that IRRGA converges much faster than the Simple GA, which does not have any redundancy in the strings. Superior effectiveness of IRRGA has been demonstrated in condition monitoring problems (Chou and Ghaboussi, 2001; Limsamphancharon, 2003).

8.4 UNSTRUCTURED DESIGN OF A PLANE TRUSS

In this section, we will present an example of an unstructured design problem and a special form of dynamic variable allocation genetic algorithm to solve this unstructured design problem. This example was part of the doctoral dissertation of Dr. Shrestha (Shrestha, 1998; Shrestha and Ghaboussi, 1998).

8.4.1 Problem definition

The objective is to design a truss to span a fixed distance. The truss must be designed to satisfy all the applicable requirements of the design code. All the member stresses, both in tension and compression, must not exceed the allowable stress limits and the displacements must remain within predefined bounds. For the members in compression, the allowable stresses are reduced to account for buckling. We will assume that the members of the truss can only be chosen from a set of standard sections. Of course, the design must be achieved with the minimum amount of material, or it must be a minimum weight design.

The choices may appear obvious to an experienced engineer. Over the years, a number of truss configurations and topologies have been designed and constructed. These trusses have been designed and evaluated many times, and they form the experience base, which is utilized

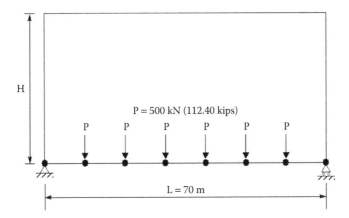

Figure 8.6 The physical design space and the loads for the truss design. (From Shrestha, S.M., Genetic algorithm based methodology for unstructured design of skeletal structures, PhD thesis, Department of Civil and Environmental Engineering, University of Illinois at Urbana-Champaign, Urbana, IL, 1998.)

in the design of new truss structures. An unstructured formulation of the problem must allow the emergence of the new designs independent of these or any other designs used in the past.

To allow the emergence of the design free of any preconceptions, only the physical space for the truss is specified. As shown in Figure 8.6, the physical design space is a rectangular region of space over the span to contain the truss. This is the only restriction being imposed on the topology of the truss, and this is not a severe restriction since most trusses are likely to lie above the span. The height of the physical design space is likely to influence the design. We will present results for two different heights of the physical design space.

8.4.2 String representation and structural synthesis

The topology of the truss structure is defined by the nodal points and the structural elements that connect those nodes. Both the structural nodes and elements are encoded in the genetic strings in a dynamic variable allocation style.

The genetic string contains one substring per node and the total number of the nodal substrings is equal to the estimated maximum possible number of nodes. The nodal coordinates are encoded in the nodal substrings so that the nodal points can move within the physical design space. An activity indicator is also encoded in the nodal substrings to indicate whether the node is active or inactive. Only the active nodes are part of the truss topology. The dynamic variable allocation is implemented through nodal activity and member activity, as will be described later. The nodes can appear and disappear depending on the value of the activity indicator.

Around each node, a number of sectors are defined, as shown in Figure 8.7. Each potential member will fall within a nodal sector at each end. The nodal sectors carry the activity indicator, priority indicator and member properties. A member is active if the activity indicators at both of its end sectors are on. The priority indicator determines which sector is used for the member properties. The member properties are not fixed, they are encoded in the nodal substrings and, as such, they can change during the evolution. In this example, the material properties are encoded as material type. This allows the members to be selected from a fixed number of predefined member types.

In the representation scheme used for this problem, only the nodes are explicitly encoded in the genetic string. The members are not encoded explicitly. They emerge from the nodal

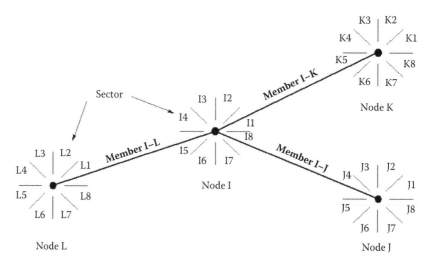

Figure 8.7 Nodal sector scheme and member representation design. (From Shrestha, S.M., Genetic algorithm based methodology for unstructured design of skeletal structures, PhD thesis, Department of Civil and Environmental Engineering, University of Illinois at Urbana-Champaign, Urbana, IL, 1998.)

sectors. It is important to note that with this type of dynamic variable allocation, there are large segments of the genetic string that are redundant at any given time. All the nodal substrings fall into this category, in the sense that they do have large redundant segments (Figure 8.8).

A total of 40 substrings were used corresponding to potential maximum of 40 nodes. Around each node 20 equal sectors were assumed.

8.4.3 Fitness function

The fitness function is defined to create selective pressure towards minimizing the cost of the structure while at the same time satisfying a number of constraints. The cost of the structure is made up of the cost of materials, labor, manufacturing of the components, and erection of the structure. In this simple example, only the cost of materials was used and the objective function C, to be minimized, is defined as the total volume of the materials. The constraints are imposed through penalty functions. Constraints are imposed on stresses, slenderness ratios, nodal displacements, minimum member lengths, maximum member lengths, nodal symmetry, and member symmetry.

The same form of penalty function is used for all the constraints. The penalty incurred by the structure for violating each constraint P_k is the product of penalty incurred by all the components of the structure for violating that constraint, as shown in the following equation:

$$P_k = \prod_{j=1}^{N_{Ck}} \left(1 + \frac{v_{kj}}{V_k}\right) \tag{8.1}$$

where:

v_{kj} is the measure of violation of constraint k by component j of the structure (j refers to a node or a member),

V_k is the constant corresponding to constraint k,

N_{Ck} is the number of components of the structure over which constraint k is applicable

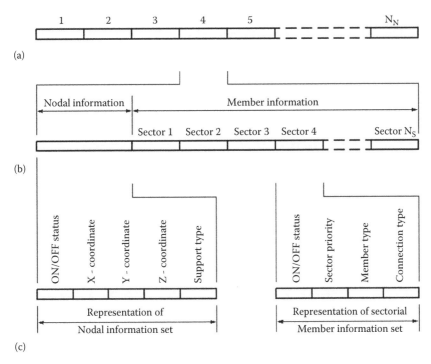

Figure 8.8 String representation and nodal and member data design: (a) string representing a structure, composed of N_N substrings; (b) structure of substring corresponding to a specific node; (c) representation of nodal and member information. (From Shrestha, S.M., Genetic algorithm based methodology for unstructured design of skeletal structures, PhD thesis, Department of Civil and Environmental Engineering, University of Illinois at Urbana-Champaign, Urbana, IL, 1998.)

The expression within the parentheses will be equal to one if the constraint is satisfied and greater than one if the constraint is violated. Its value is the measure of the violation of the constraint. The value of v_{kj} is zero when the constraint is satisfied and greater zero when the constraint is violated.

As an example, we will consider the stress constraint, which is expressed by the following equation.

$$\frac{\sigma_j}{\sigma_{aj}} \leq 1 \qquad (8.2)$$

where:
  σ_j is the stress in member j,
  σ_{aj} is the allowable stress in tension or the allowable stress in compression, adjusted for buckling considerations

The terms of the penalty function for this constraint are given in the following equations:

$$v_{kj} = w_j \left(\frac{\sigma_j}{\sigma_{aj}} - 1 \right) \qquad (8.3)$$

$$\begin{cases} V_k = \sum_{j=1}^{N_m} w_j \\ N_{Ck} = N_m = \text{number of members} \end{cases} \tag{8.4}$$

where w_j is the weight of member j. Similar expressions are used for the penalty functions corresponding to the other constraints (Shrestha, 1998; Shrestha and Ghaboussi, 1998).

The total cost function, C^T, is determined from the objective function and the weighted penalty functions:

$$C^T = C \prod_{k=1}^{N_k} (P_k)^{a_k} \tag{8.5}$$

where:
   C is the objective function expressed as the cost function that needs to be minimized,
   N_K is the number of constraints, and
   a_k is the exponential penalty weight factor for constraint k.

The fitness function is related to the inverse of the total cost function. First, a raw fitness function is calculated according to the following equation:

$$f' = \frac{C^T_{max}}{C^T} \tag{8.6}$$

where C^T_{max} is the total cost of the least fit structure in the current generation. It can be seen that the raw fitness of all the structures are greater than 1.0 and the raw fitness of the least fit structure is equals 1.0. The actual fitnesses are determined from the raw fitnesses through a special scaling. The details of this scaling can be found in Shrestha (1998), Shrestha and Ghaboussi (1998).

8.4.4 Evolution of a truss design

The methodology described in the previous section was applied to the design of two single span trusses. The span in both cases is 70 m, and the loading is shown in Figure 8.6. In the first case, the height of the physical design space is 35 m. In both cases, the strings consisted of 40 substrings, representing a maximum of 40 nodes. The strings were 25,200-bit long. In the initial population, strings were generated randomly. The history of the evolution of the truss is shown in Figure 8.9, which shows the fittest member at several generations. The total cost of the fittest member in each generation is shown in Figure 8.10 for 10,000 generations.

The process of arriving at a design and optimizing that design is clear. Figure 8.10 shows three distinct phases. The first phase, approximately the first 3,500 generations, is characterized by extreme oscillations in the cost profile. Different designs are being explored in this phase. The second phase begins when a potentially promising design has emerged. This phase begins with a sharp drop in the cost profile, and it ends when the cost function reaches a plateau. Major changes in the topology and the configuration of the structure do occur in the second phase. Some nodes are eliminated, and other nodes are repositioned to yield better member sizes and member alignments. The cost function is reasonably flat during the final phase, which is characterized by a gradual improvement in the structural design. All redundant nodes and members disappear. The primary emphasis in this phase appears to be in the reduction in weight and on satisfying the constraints. The basic design remains

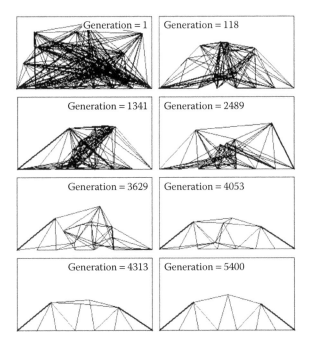

Figure 8.9 History of the evolution of truss design—The fittest member in each generation is shown. (From Shrestha, S.M., Genetic algorithm based methodology for unstructured design of skeletal structures, PhD thesis, Department of Civil and Environmental Engineering, University of Illinois at Urbana-Champaign, Urbana, IL, 1998; Shrestha, S.M. and Ghaboussi, J., *J. Struct. Eng.*, 124, 1331–1338, 1998.)

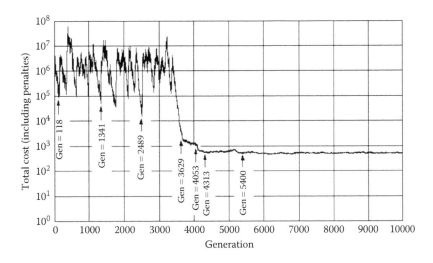

Figure 8.10 The cost of the fittest member in each generation during the evolution of the truss design. (From Shrestha, S.M., Genetic algorithm based methodology for unstructured design of skeletal structures, PhD thesis, Department of Civil and Environmental Engineering, University of Illinois at Urbana-Champaign, Urbana, IL, 1998; Shrestha, S.M. and Ghaboussi, J., *J. Struct. Eng.*, 124, 1331–1338, 1998.)

unaltered in later stages of the third phase, and the changes in the structure are in the nature of fine tuning of the design. At this stage, genetic algorithm is performing a task that is similar to optimization. Once an optimal design has been discovered in first two phases, the methodology optimizes it in the third phase.

Significant changes in the topology and configuration from that shown in Figure 8.9 at generation 5,400 would cause a sharp decrease in the fitness function and it will be quickly selected out.

It is interesting to note that the truss that has evolved is a statically determinate truss. The nodal and member symmetry constraints have been satisfied reasonably well through the penalty functions, since the resulting structure is reasonably symmetric. The resulting structure will not be exactly symmetric, since the basic methodology is imprecision tolerant; the difference between the approximate symmetry and the perfect mathematical symmetry is too small to cause any significant selection pressure.

The final design shown Figure 8.9 can also be considered an innovative and creative design. This design does not fit well into the designs that exist in the experience base of most structural designers, and it is not a design that is familiar to most designers. The author was pleasantly surprised when he saw this design for the first time.

Figures 8.11 and 8.12 show the evolution of a different truss for the same span and same loading as in the previous design. The only difference is that the height of the physical design space in this case is smaller; it is 10 m. In the previous case, the height of the physical design space was 35 m, and the evolved designs only needed half that height. In this case, the designs evolve to utilize the full height of the physical design space. Since the height of the truss is smaller, the top and the bottom chord members need to carry larger loads than in the previous case. It is known that the shorter members are more efficient in carrying large compressive loads, and genetic algorithm has evolved short members for the top chord of the truss.

The three phases of the evolution of the truss are evident Figure 8.12, although the first phase is far shorter in this case.

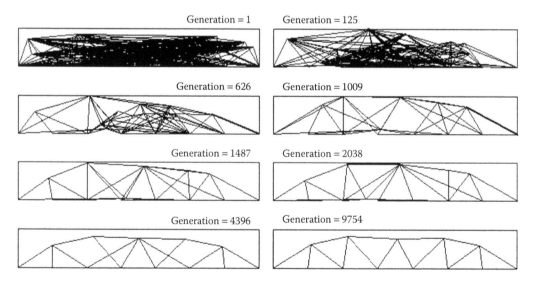

Figure 8.11 History of the evolution of truss design in a shallow physical design space. The fittest member in each generation is shown. (From Shrestha, S.M., Genetic algorithm based methodology for unstructured design of skeletal structures, PhD thesis, Department of Civil and Environmental Engineering, University of Illinois at Urbana-Champaign, Urbana, IL, 1998; Shrestha, S.M. and Ghaboussi, J., *J. Struct. Eng.*, 124, 1331–1338, 1998.)

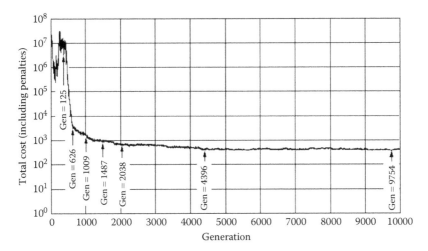

Figure 8.12 The cost of the fittest member in each generation during the evolution of the truss design in a shallow physical design space. (From Shrestha, S.M., Genetic algorithm based methodology for unstructured design of skeletal structures, PhD thesis, Department of Civil and Environmental Engineering, University of Illinois at Urbana-Champaign, Urbana, IL, 1998; Shrestha, S.M. and Ghaboussi, J., *J. Struct. Eng.*, 124, 1331–1338, 1998.)

8.5 IRRGA IN UNSTRUCTURED DESIGN OF A PLANE FRAME

Earlier in this chapter and also in Chapter 7, we stated that autogenesis and redundancy make IRRGA ideal for unstructured design problems. In this section, we present an example of such an application. This example was part of Dr. Anne Raich's doctoral dissertation (Raich, 1998; Raich and Ghaboussi, 1997, 2000a, 2000b, 2001).

The example involves the design of a plane frame structure, used as a part of the skeletal frame in a building. The loads to be carried by the plane frame are specified as the dead load, live load, and wind load. The structure is designed according to the design code and must satisfy all its provisions. It is also specified that building, and the plane frame, has three stories.

In designing such a three-story plane frame, one often does not think of creativity. The conventional design calls for horizontal girders at the floor levels and vertical columns. The horizontal wind load is either resisted by designing the structure to resist the lateral loads as a moment-resisting frame (the connection can transfer moment), or by using diagonal bracing elements. It is also possible to use both methods. In this example, we assume that the frame is moment resisting.

The physical domain for the design synthesis of the three-story plane frame and the loads are shown in Figure 8.13. The floor levels are defined on the basis of the floor heights that have more or less standard dimensions. The plane frame is to be designed within the physical domain, with the specified floor levels and to carry the loads shown and to satisfy the provision of the design code.

Standard designs for the three-story plane frames are very simple. They consist of horizontal girders, vertical columns, and occasionally diagonal bracing members. Three examples of the standard designs are shown in Figure 8.14.

In the unstructured design format, we only wish to predefine the physical domain that the frame should occupy, the three floor levels, the loads, and the provisions of the design code. Nothing else is specified or predefined. The horizontal girders are needed at the floor levels,

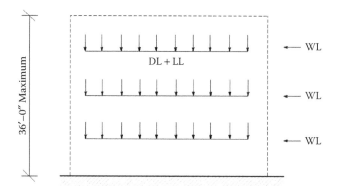

Figure 8.13 The unstructured problem domain for the plane frame design. (From Raich, A.M., An evolutionary based methodology for representing and evolving structural design solutions, PhD thesis, Department of Civil and Environmental Engineering, University of Illinois at Urbana-Champaign, Urbana, IL, 1998; Raich, A.M. and Ghaboussi, J., *Int. J. Evol. Comput.*, 5, 277–302, 1997; Raich, A.M. and Ghaboussi, J., *Practical Handbook of Genetic Algorithm: Applications*, Chambers, L. (Ed.), Chapman & Hall/CRC publishers, Boca Raton, FL, 2000a; Raich, A.M. and Ghaboussi, J., *J. Struct. MultiDiscip. O.*, 20, 222–231, 2000b; Raich, A.M. and Ghaboussi, J., *Struct. O.*, 20, 222–223, 2001.)

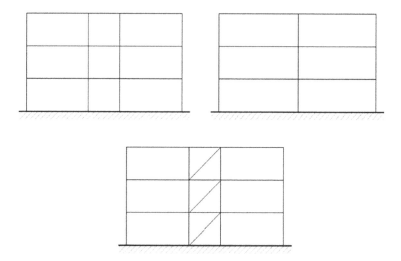

Figure 8.14 Three examples of the standard design of the three-story plane frame.

but the spans of those girders are not specified. No other members are predefined. Specially, we do not require that the frame have vertical columns.

The main purpose of formulating the design of the three-story plane frame as an unstructured design problem is to determine whether innovative and creative new designs can emerge through the evolution. The details of the formulation of the problem can be found in the doctoral dissertation of Dr. Anne Raich (Raich, 1998) and in Raich and Ghaboussi (1997, 2000a, 2000b, 2001). Here, we will describe the formulation and present and discuss some of the results.

The design variables for defining the nonhorizontal member coordinates, nodal incidences, and member depths are shown in Figure 8.15. These design variables for each nonhorizontal member are encoded in a single gene instance identified by the Gene Locator Patter (GLP) [1 1 1]. The order of the variables in the gene instance is shown in Figure 8.16. Horizontal

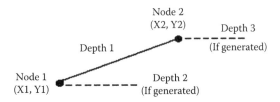

Figure 8.15 Design variables for one member in the unstructured design of plane frame encoded the gene instances of IRRGA. (From Raich, A.M., An evolutionary based methodology for representing and evolving structural design solutions, PhD thesis, Department of Civil and Environmental Engineering, University of Illinois at Urbana-Champaign, Urbana, IL, 1998; Raich, A.M. and Ghaboussi, J., *Int. J. Evol. Comput.*, 5, 277–302, 1997; Raich, A.M. and Ghaboussi, J., *Practical Handbook of Genetic Algorithm: Applications*, Chambers, L. (Ed.), Chapman & Hall/CRC publishers, Boca Raton, FL, 2000a; Raich, A.M. and Ghaboussi, J., *J. Struct. MultiDiscip. O.*, 20, 222–231, 2000b; Raich, A.M. and Ghaboussi, J., *Struct. O.*, 20, 222–223, 2001.)

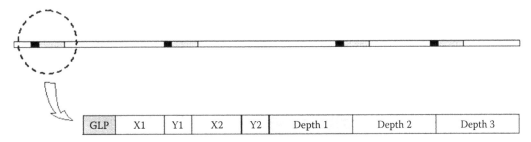

Figure 8.16 Segment of IRRGA string and the gene instance for the unstructured frame design problem. (From Raich, A.M., An evolutionary based methodology for representing and evolving structural design solutions, PhD thesis, Department of Civil and Environmental Engineering, University of Illinois at Urbana-Champaign, Urbana, IL, 1998; Raich, A.M. and Ghaboussi, J., *Int. J. Evol. Comput.*, 5, 277–302, 1997; Raich, A.M. and Ghaboussi, J., *Practical Handbook of Genetic Algorithm: Applications*, Chambers, L. (Ed.), Chapman & Hall/CRC publishers, Boca Raton, FL, 2000a; Raich, A.M. and Ghaboussi, J., *J. Struct. MultiDiscip. O.*, 20, 222–231, 2000b; Raich, A.M. and Ghaboussi, J., *Struct. O.*, 20, 222–223, 2001.)

decoded members with the same Y coordinates at both ends are ignored in assembling the frame. The total number of members has not been specified, and they will vary between the strings in the population.

The design variable value ranges are set by the number of binary bits used to encode each variable. The nodal X-coordinates, X1, and X2, are encoded as 6-bit binary numbers that are mapped by the following function: (X1–31.0) * 12.0, which encodes a value range of (−372.0, 384.0) in inches. The Y-coordinates, Y1 and Y2, are encoded as 2-bit binary numbers. Each of the four encoded binary values corresponds to a floor level of 0, 1, 2, or 3. All three member cross-sectional depths are 3-bit binary numbers that encode eight discrete member depths {5, 10, 15, 20, 25, 30, 40, 50} with a unit of inches. All the structural frame members are defined as steel tube sections having a fixed width and thickness and a variable decoded depth. The member area and the section modulus are calculated on the basis of the member depth decoded.

The two horizontal member cross-sectional depths decoded from the gene instance for each nonhorizontal member are used when a horizontal member is generated. The horizontal members are generated between each pair of adjacent nodes defined on the same floor level from the nonhorizontal member information decoded. The depth of horizontal member is provided by the value of the horizontal cross-sectional depth (Depth 2 or Depth 3) decoded for the designated starting node of the horizontal member as shown in Figure 8.15. Assembling a complete frame structure consists of defining the nonhorizontal member locations using the nodal coordinates decoded from the IRR genetic string and generating the horizontal members by connecting the nodal coordinates defined along each level.

The decoded structures need to be analyzed to determine the displacements and the maximum member stresses for evaluating the fitness of each individual. It is possible that the decoded structures may not be stable. Repair strategies need to be defined to make the structural analysis of those structures possible. The repair strategies used in this example are described in Raich (1998), Raich and Ghaboussi (1997, 2000a, 2000b, 2001). The important consideration in devising a repair strategy is to create a stable structure to enable structural analysis without artificially increasing its fitness. Often, this is accomplished by adding members with very small section properties. Although such a structure is theoretically stable and can be analyzed, it will have very large displacements and very low fitness value. These structures with very low fitness values are likely to be quickly selected out. The reason that they are not eliminated outright is that they may have other desirable trait, and they will have a small chance of passing those traits to the future generations.

The details of the fitness function evaluation are also given in Raich (1998), Raich and Ghaboussi (1997, 2000a, 2000b, 2001). For the purpose of this book, it suffices to state the main step in fitness evaluation. Once a structure is decoded, it is examined to verify its stability. If the structure is not stable, then the predefined retrofit strategies are applied to stabilize the structure. Then, the structure is analyzed under the loading combinations specified by the Load Resistant Factor Design code, and the displacements of the nodes and stresses in the members are determined. The fitness function is formulated to minimize the total weight of the structure, while maximizing the total floor area. The fitness function also has terms for penalizing excessive horizontal and vertical displacements, member stresses in excess of allowable stresses, and the lack of symmetry.

The string length is an important consideration in IRRGA; it controls the redundancy ratio and the number of gene instances in the randomized initial population. The string length can strongly influence the performance of the IRRGA. If it is too short to allow the emergence of sufficient number of variables, it may hinder or prevent convergence to an admissible optimal solution. In this example, a string length of 600 bit was deemed to be adequate.

Figure 8.17 shows the evolution of one design. It is important to note that many acceptable solutions are possible. Each application of IRRGA, starting from random initial conditions, may yield a different solution. The creative aspects of the evolved solution will be discussed later. At this point, we study the process of the emergence of the admissible solution. Shown in Figure 8.17 is the fittest member in each generation. The first generation is randomly created and as a result, it is not a very good design. The outline of a good and admissible solution appears in generation number 50, and the synthesis of the frame is complete by generation number 200. Between generation numbers 200 and 500 IRRGA is optimizing the frame members, which is a classical optimization problem. The two stages of design synthesis and optimization are marked in Figure 8.18.

This boundary between the synthesis and optimization shown in Figure 8.18 is approximate, and it is not as sharply delineated as shown in the figure. The transition from synthesis to optimization is very gradual. Once the method finds a "good" admissible solution, like the configuration in generation number 200, it is not likely to change, even though the low average fitness for the population indicates that there is considerable genetic diversity in the population. The increases in the fitness function from this point on are the result of the method optimizing the section properties to reduce the total volume of the material and to reduce the excessive displacements and member stresses.

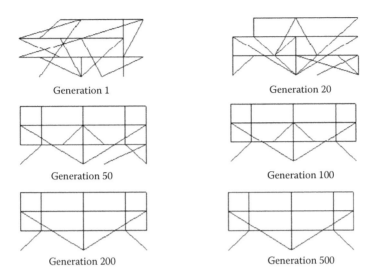

Figure 8.17 An example of the evolution of the best IRRGA design solution at each generation. (From Raich, A.M., An evolutionary based methodology for representing and evolving structural design solutions, PhD thesis, Department of Civil and Environmental Engineering, University of Illinois at Urbana-Champaign, Urbana, IL, 1998; Raich, A.M. and Ghaboussi, J., *Int. J. Evol. Comput.*, 5, 277–302, 1997; Raich, A.M. and Ghaboussi, J., *Practical Handbook of Genetic Algorithm: Applications*, Chambers, L. (Ed.), Chapman & Hall/CRC publishers, Boca Raton, FL, 2000a; Raich, A.M. and Ghaboussi, J., *J. Struct. MultiDiscip. O.*, 20, 222–231, 2000b; Raich, A.M. and Ghaboussi, J., *Struct. O.*, 20, 222–223, 2001.)

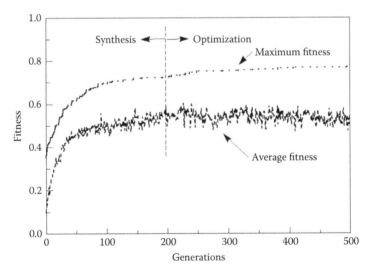

Figure 8.18 Maximum and average fitness of the IRRGA population over 500 generations for a single trial with a 10'–0" restriction on the x-coordinate spacing. (From Raich, A.M., An evolutionary based methodology for representing and evolving structural design solutions, PhD thesis, Department of Civil and Environmental Engineering, University of Illinois at Urbana-Champaign, Urbana, IL, 1998; Raich, A.M. and Ghaboussi, J., *Int. J. Evol. Comput.*, 5, 277–302, 1997; Raich, A.M. and Ghaboussi, J., *Practical Handbook of Genetic Algorithm: Applications*, Chambers, L. (Ed.), Chapman & Hall/CRC publishers, Boca Raton, FL, 2000a; Raich, A.M. and Ghaboussi, J., *J. Struct. MultiDiscip. O.*, 20, 222–231, 2000b; Raich, A.M. and Ghaboussi, J., *Struct. O.*, 20, 222–223, 2001.)

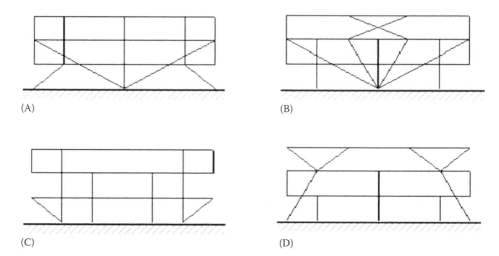

(A)   (B)

(C)   (D)

Figure 8.19 IRRGA frame design solutions represented by the fittest population individual after 500 generations. (From Raich, A.M., An evolutionary based methodology for representing and evolving
structural design solutions, PhD thesis, Department of Civil and Environmental Engineering,
University of Illinois at Urbana-Champaign, Urbana, IL, 1998; Raich, A.M. and Ghaboussi, J.,
Int. J. Evol. Comput., 5, 277–302, 1997; Raich, A.M. and Ghaboussi, J., *Practical Handbook of Genetic
Algorithm: Applications*, Chambers, L. (Ed.), Chapman & Hall/CRC publishers, Boca Raton, FL,
2000a; Raich, A.M. and Ghaboussi, J., *J. Struct. MultiDiscip. O.*, 20, 222–231, 2000b; Raich, A.M.
and Ghaboussi, J., *Struct. O.*, 20, 222–223, 2001.)

Figure 8.19 shows four frames synthesized in four separate applications of IRRGA, each
starting from a randomized initial population. After a careful study of these frames, the
reader is likely to agree that these are innovative and creative designs. It is an indication
that, when the problem is formulated as an unstructured design problem, IRRGA is able to
produce unexpectedly creative designs even in the case of a simple three-story plane frame.

With the exception of frame C, none of these frames carry the vertical loads through vertical column members alone. They have many inclined members that combine the carrying
of the vertical loads with resisting the lateral wind loads. The basic theme through most of
these frames is the fact that the structural members are far more efficient in carrying axial
load than bending moments. We can clearly see this in frames A, B, and D. The synthesis has
relied mostly on the use of the trusses and arches as the main load carrying elements in the
structure. Both of these structural types support the loads mainly by axial forces.

In frame A, the three floors are suspended from a large inverted triangular truss at the base of
the structure. Triangular trusses are very efficient in carrying the load. The two inclined members
on the first floor provide the lateral stability for the structure. Similar pattern can be seen in frame B,
in which two triangular trusses have been deployed in the first floor of the structure.

In frame D, the main load carrying element in the frame is an arch. Since the arch is made
up of straight line members, it may have a tendency to be unstable. To further stabilize the
arch, two triangular elements have been introduced in the top floor of the frame. The two
lower floors have been suspended from the arch. The three short vertical columns are introduced to reduce the spans of the horizontal girders.

In summary, these four designs have creative aspects. They have emerged through the evolutionary process with the minimum of interference from the outside the method. We did not
have to specify the topology of the structure, nor the number and coordinate of the nodes, nor
the member connectivity. This was possible mainly because the problem was formulated as
an unstructured design problem and a method with autogenesis and redundancy were used.

8.6 SUMMARY AND DISCUSSION

The number of variables in a vast majority of optimization problems is fixed. Classical mathematically based optimization methods as well as the SGA can be used to solve most of these optimization problems. On the other hand, the number and the identity of the variables in a design problem are not known at the outset. We have discussed the fact that, while the optimization is a search in finite-dimensional vector space, design is a search in the infinite-dimensional vector space. The term "unstructured design problems" was used to describe problems that lie somewhere between optimization and design problems. In the unstructured design problems, some design decisions have already been made and the remaining part of the problem is similar to design problems, in that the number and identity of the variables are not known at the outset. Another attribute they share with the design problems is the considerable scope for creativity and innovation, unlike the optimization problems that offer limited scope for creativity.

Genetic algorithm has been successfully used in optimization problems. The primary objective of this chapter is to show that genetic algorithm has capabilities beyond the optimization problems and that it can be used in solving the difficult design problems. Static variable allocation in the SGA cannot be used in design problems, since the number and identity of the variables are not known a-priory. We introduced the concept of dynamic variable allocation, which allows the appearance of new variables and disappearance of existing variables during the evolution. We introduced two different methods of implementing dynamic variable allocation.

The first method of dynamic variable allocation differs from the static variable allocation in that each variable or substring has an on/off switch indicating its status. Only the variables that have their on/off switch on are decoded to describe the design, whereas those with their switch off are ignored and they become redundant parts of the string. Of course, the redundant parts can become active in future generations if their on/off switch turns on.

The second method is the Implicit Redundant Representation, described in Chapter 7, where encountering a predefined sequence of bits called gene locator pattern (GLP), like [1, 1, 1], during the parsing of the strings indicates the existence of a substring immediately following GLP sequence.

Both methods of dynamic variable allocation allow segments of the strings to remain redundant and unused at any time during the evolution. The redundant segments are also dynamic and they change during the evolution.

Two examples were presented to illustrate the application of the two methods of dynamic variable allocation in unstructured design problems. In the first example, the objective was to design a truss within a physical design space to carry a specified load. The topology and the configuration of the truss was part of the design problem and it was not specified. The method of dynamic variable allocation with the on/off switches was used in this example. The design yielded a fully stressed and statically determinate truss that satisfied the allowable stress design requirements, including the buckling considerations, and the symmetry conditions that were enforced through penalty functions.

In the second example, the IRRGA was used to design a three-story moment resisting plane frame within a physical design space to carry the dead, live and wind loads and to have the maximum floor space while having the minimum weight. Constraints were imposed through penalty functions to satisfy the allowable stress design code requirements and symmetry conditions.

The resulting designs in both examples had creative features, indicating that creative designs can emerge during the evolution when genetic algorithm is formulated to allow dynamic variable allocation and redundancy.

The experience based engineering design in practice goes through two main stages of conceptual design and final design. The design is synthesized during the conceptual design which is an open-ended process with considerable scope for creativity. The final design stage is similar to an optimization problem with a fixed number of variables. The two examples showed that the in the unstructured design problems, genetic algorithm goes through two distinctly identifiable stages of synthesize and optimization.

The emphasis in this chapter has been mainly on the application of genetic algorithm with dynamic variable allocation in engineering design. We should point out the methodology has potential applications beyond the engineering design; for example, in fields such as architectural design and urban development. An example of application of the IRRGA in architectural design is presented in Song et al. (2016). It uses IRRGA in the conceptual design of apartment buildings, mainly from architectural point of view. The location, size, orientation, and number of the apartment units are the main variables. The rules enforced through the fitness function are mainly architectural rules, such as aesthetics, symmetry, connectivity and access to the apartments units, structural feasibility and total cost. The results show a degree of creativity in the design of apartment buildings.

References

Aboudi, J. (1991) *Mechanics of Composite Materials-A Unified Micromechanical Approach*, Amsterdam, the Netherlands: Elsevier.

Bani-Hani, K. and Ghaboussi, J. (1998a) Nonlinear structural control using neural networks, *Journal of Engineering Mechanics Division, ASCE*, 124(3): 319–327.

Bani-Hani, K. and Ghaboussi, J. (1998b) Neural networks for structural control of a benchmark problem, active tendon system, *International Journal for Earthquake Engineering and Structural Dynamics*, 27: 1225–1245.

Bani-Hani, K., Ghaboussi, J. and Schneider, S. P. (1999a) Experimental study of identification and structural control using neural network: I. identification, *International Journal for Earthquake Engineering and Structural Dynamics*, 28: 995–1018.

Bani-Hani, K., Ghaboussi, J. and Schneider, S. P. (1999b) Experimental study of identification and structural control using neural network: II. control, *International Journal for Earthquake Engineering and Structural Dynamics*, 28: 1019–1039.

Chang, F-K. and Lessard, L. (1991) Damage tolerance of laminated composites containing an open hole and subjected to compressive loadings: Part I-analysis, *Journal of Composite Materials*, 25: 2–43.

Chou, J. H. (2000) Study of condition monitoring of bridges using genetic algorithms, PhD thesis, Department of Civil and Environmental Engineering, University of Illinois at Urbana-Champaign, Urbana, IL.

Chou, J. H. and Ghaboussi, J. (2001) Genetic algorithm in structural damage detection, *International Journal of Computers and Structures*, 79: 1335–1353.

Elnashai, A. S., Elghazouli, A. Y. and Denesh-Ashtiani, F. A. (1998) Response of semirigid steel frames to cyclic and earthquake loads, *Journal of Structural Engineering, ASCE*, 124(8): 857–867.

Ghaboussi, J. (2001) Biologically inspired soft computing methods in structural mechanics and engineering, *International Journal of Structural Engineering and Mechanics*, 11(5): 485–502.

Ghaboussi, J., Garrett, J. H. and Wu, X. (1990) Material modeling with neural networks, *Proceedings, NUMETA-90, International Conference on Numerical Methods in Engineering: Theory and Applications*, Swansea, England.

Ghaboussi, J., Garrett, J. H. and Wu, X. (1991) Knowledge-based modeling of material behavior with neural networks, *Journal of Engineering Mechanics Division, ASCE*, 117(1): 132–153.

Ghaboussi, J. and Insana, M. F. (2017) *Understanding Systems: A Grand Challenge for 21st Century Engineering*, Singapore: World Scientific Publishing.

Ghaboussi, J. and Joghataie, A. (1995) Active control of structures using neural networks, *Journal of Engineering Mechanics Division, ASCE*, 121(4): 555–567.

Ghaboussi, J., Kwon, T.-H., Pecknold, D. A. and Hashash, Y. M. A. (2009) Accurate intraocular pressure prediction from applanation response data using genetic algorithm and neural networks, *Journal of Biomechanics*, 42(14): 2301–2306.

Ghaboussi, J., Lade, P. V. and Sidarta, D. E. (1994) Neural network based modeling in geomechanics, *Proceedings, International Conference on Numerical Methods and Advances in Geomechanics*.

Ghaboussi, J. and Lin, C.-C. J. (1998) A new method of generating earthquake accelerograms using neural networks, *International Journal for Earthquake Engineering and Structural Dynamics*, 27: 377–396.

Ghaboussi, J., Pecknold, D. A. and Wu, X. S. (2017) *Nonlinear Computational Solid Mechanics*, Boca Raton, FL: CRC Press.

Ghaboussi, J., Pecknold, D. A., Zhang, M. and HajAli, R. (1998) Autoprogressive training of neural network constitutive models, *International Journal for Numerical Methods in Engineering*, 42: 105–126.

Ghaboussi, J. and Sidarta, D. E. (1998) A new nested adaptive neural network for modeling of constitutive behavior of materials, *International Journal of Computer and Geotechnics*, 22(1): 29–51.

Ghaboussi, J. and Wu, X. (1998) Soft computing with neural networks for engineering applications: Fundamental issues and adaptive approaches, *International Journal of Structural Engineering and Mechanics*, 6(8): 955–969.

Ghaboussi, J. and Wu, X. S. (2016) *Numerical Methods in Computational Mechanics*, Boca Raton, FL: CRC Press.

Goldberg, D. E. (1989) *Genetic Algorithms in Search, Optimization, and Machine Learning*, Boston, MA: Addison-Wesley.

Hashash, Y. M. A., Fu, Q.-W., Ghaboussi, J., Lade, P. V. and Saucier, C. (2009) Inverse analysis based interpretation of sand behavior from triaxial shear tests subjected to full end restraint, *Canadian Geotechnical Journal*, 46(7): 768–791.

Hashash, Y. M. A., Jung, S. and Ghaboussi, J. (2004) Numerical implementation of a neural networks based material model in finite element, *International Journal for Numerical Methods in Engineering*, 59: 989–1005.

Hashash, Y. M. A., Song, H., Jung, S. and Ghaboussi, J. (2009) Extracting inelastic metal behavior through inverse analysis: A shift in focus from material models to material behavior, *Inverse Problems in Science and Engineering*, 17(1): 35–50.

Hoerig, C., Ghaboussi, J. and Insana, M. F. (2016) An information-based machine learning approach to elasticity imagining, *Biomechanics and Modeling in Mechanobiology*, 16(3): 805–822.

Hoerig, C., Ghaboussi, J. and Insana, M. F. (2017) A machine learning alternative to model-based elastography, *The Journal of the Acoustical Society of America*, 141(5): 3675.

Holland, J. (1975) *Adaptation in Natural and Artificial Systems*, Cambridge, MA: The MIT Press.

Hopfield, J. J. (1982) Neural networks and physical systems with emergent collective computational abilities, *Proceedings of the National Academy of Sciences of the USA*, 79(8): 2554–2558.

Joghataie, A. (1994) Neural networks and fuzzy logic for structural control, PhD thesis, Department of Civil and Environmental Engineering, University of Illinois at Urbana-Champaign, Urbana, IL.

Joghataie, A., Ghaboussi, J. and Wu, X. (1995) Learning and architecture determination through automatic node generation, *Proceedings, International Conference on Artificial neural Networks in Engineering*, St Louis, MO.

Jung, S.-M. (2004) Field calibration and analysis of creep in prestressed concrete segmental bridges, PhD thesis, Department of Civil and Environmental Engineering, University of Illinois at Urbana-Champaign, Urbana, IL.

Jung, S.-M. and Ghaboussi, J. (2006a) Neural network constitutive model for rate-dependent materials, *Computer and structures*, 84: 955–963.

Jung, S.-M. and Ghaboussi, J. (2006b) Characterizing rate-dependent material behaviors in self-learning simulation, *Computer Methods in Applied Mechanics and Engineering*, 196(1–3): 608–619.

Jung, S.-M. and Ghaboussi, J. (2010) Inverse identification of creep of concrete from in-situ load-displacement monitoring, *Journal of Engineering Structures*, 32(5): 1437–1445.

Jung, S.-M., Ghaboussi, J. and Marulanda, C. (2007) Field calibration of time-dependent behavior in segmental bridges using self-learning simulations, *Engineering Structures*, 29(10): 2692–2700.

Karsan, I. D. and Jirsa, J. O. (1969). Behavior of concrete under compressive loading, *Journal of the Structural Division, ASCE*, 95: 2543–2563.

Kim, J-H. (2009) Hybrid physical and informational modeling of beam-column connections. PhD thesis, Department of Civil and Environmental Engineering, University of Illinois at Urbana-Champaign, Urbana, IL.

Kim, J.-H., Ghaboussi, J. and Elnashai, A. S. (2010) Mechanical and informational modeling of steel beam-column connections, *Journal of Engineering Structure*, 32(2): 449–458.

Kim, J.-H., Ghaboussi, J. and Elnashai, A. S. (2012) Hysteretic mechanical-informational modeling of bolted steel frame connections, *Engineering Structure*, 45(1): 1–11.

Kwon, T.-H. (2006) Minimally invasive characterization and intraocular pressure measurement via numerical simulation of human cornea. PhD thesis, Department of Civil and Environmental Engineering, University of Illinois at Urbana-Champaign, Urbana, IL.

Kwon, T.-H., Ghaboussi, J., Pecknold, D. A., and Hashash, Y. M. A. (2008) Effect of cornea material stiffness on measured intraocular pressure, *Journal of Biomechanics*, 41(8): 1707–1713.

Kwon, T.-H., Ghaboussi, J., Pecknold, D. A. and Hashash, Y. M. A. (2010) Role of cornea biomechanical properties in applanation tonometry measurements, *Journal of Refractive Surgery*, 26(7): 512–519.

Lade, P. V., Ghaboussi, J., Inei, S. and Yamamuro, J. A. (1994) Experimental determination of constitutive behavior of soils, *Proceedings, International Conference on Numerical Methods and Advances in Geomechanics.*

Lessard, L. and Chang, F.-K. (1991) Damage tolerance of laminated composites containing an open hole and subjected to compressive loadings: Part II-experiment, *Journal of Composite Materials*, 25: 44–64.

Limsamphancharon, N. (2003) Condition monitoring of structures by using ambient dynamic responses, PhD thesis, Department of Civil and Environmental Engineering, University of Illinois at Urbana-Champaign, Urbana, IL.

Lin, C.-C. J. (1999) A neural network based methodology for generating spectrum compatible earthquake accelerograms, PhD thesis, Department of Civil and Environmental Engineering, University of Illinois at Urbana-Champaign, Urbana, IL.

Lin, C.-C. J. and Ghaboussi, J. (2001) Generating multiple spectrum compatible accelerograms using neural networks, *International Journal Earthquake Engineering and Structural Dynamics*, 30: 1021–1042.

McCulloch, W. S. and Pitts, W. (1943) A logical calculus of the ideas immanent in nervous activity, *Bulletin of Mathematical Biophysics*, 5: 115–133.

Newmark, N. M. and Hall, W. J. (1982) Earthquake spectra and design, *EERI Monograph Series*, Earthquake Engineering Research Institute.

Nikzad, K., Ghaboussi, J. and Paul, S. L. (1996) A study of actuator dynamics and delay compensation using a neuro-controller, *Journal of Engineering Mechanics Division, ASCE*, 122(10): 966–975.

Palermo, D. and Vecchio, F. J. (2003) Compression field modeling of reinforced concrete subjected to reversed loading: Formulation, *ACI Structural Journal*, 100: 616–625.

Pecknold, D. A. and Rahman, S. (1994) Micromechanics-based structural analysis of thick laminated composites, *Computers & Structures*, 51(2): 163–179.

Raich, A. M. (1998) An evolutionary based methodology for representing and evolving structural design solutions, PhD thesis, Department of Civil and Environmental Engineering, University of Illinois at Urbana-Champaign, Urbana, IL.

Raich, A. M. and Ghaboussi, J. (1997) Implicit representation in genetic algorithm using redundancy, *International Journal of Evolutionary Computing*, 5(3): 277–302.

Raich, A. M. and Ghaboussi, J. (2000a) Applying the implicit redundant representation genetic algorithm in an unstructured problem domain, Chapter 9 in *Practical Handbook of Genetic Algorithm, vol. 1. Applications*, L. Chambers (Ed.), Boca Raton, FL: Chapman & Hall/CRC publishers.

Raich, A. M. and Ghaboussi, J. (2000b) Evolving structural design solutions using an implicit redundant genetic algorithm, *Journal of Structural and Multi-Disciplinary Optimization*, 20(3): 222–231.

Raich, A. M. and Ghaboussi, J. (2001) Evolving the topology and geometry of frame structures during optimization, *Structural Optimization*, 20(3): 222–223.

Schmidhuber, J. (2015) Deep learning in neural networks: An overview, *Neural Networks*, 61: 85–117.

Schneider, S. P. and Teeraparbwong, I. (2002) Inelastic behavior of bolted flange plate connections, *Journal of Structural Engineering*, 128(4): 492–500.

Shrestha, S. M. (1998) Genetic algorithm based methodology for unstructured design of skeletal structures, PhD thesis, Department of Civil and Environmental Engineering, University of Illinois at Urbana-Champaign, Urbana, IL.

Shrestha, S. M. and Ghaboussi, J. (1998) Evolution of optimal structural shapes using genetic algorithm, *Journal of Structural Engineering, ASCE*, 124(8): 1331–1338.

Sidarta, D. E. (2000) Neural network based constitutive modeling of granular materials, PhD thesis, Department of Civil and Environmental Engineering, University of Illinois at Urbana-Champaign, Urbana, IL.

Sidarta, D. E. and Ghaboussi, J. (1998) Modeling constitutive behavior of materials from non-uniform material tests, *International Journal of Computer and Geotechnics*, 22(1): 53–71.

Sinha, B. P., Gerstle, K. H. and Tulin, L. G. (1964) Stress–strain relations for concrete under cyclic loading, *ACI Structural Journal*, 61: 195–211.

Smith, G. and Young, L. (1955) Ultimate theory in flexure by exponential function, *ACI Journal*, 52(3): 349–359.

Song, H., Ghaboussi, J. and Kwon, T.-H. (2016) Architectural design of apartment buildings using the implicit redundant representation genetic algorithm, *Automation in Construction*, 72(2): 166–173.

Wu, X. (1991) Neural network based material modeling, PhD thesis, Department of Civil and Environmental Engineering, University of Illinois at Urbana-Champaign, Urbana, IL.

Wu, X., Ghaboussi, J. and Garrett, J. H. (1992) Use of neural networks in detection of structural damage, *Computers and Structures*, 42(4): 649–660.

Yun, G. J., Ghaboussi, J. and Elnashai, A. S. (2007) Development of neural network based hysteretic models for steel beam-column connections through self-learning simulations, *Journal of Earthquake Engineering*, 11(3): 453–467.

Yun, G. J., Ghaboussi, J. and Elnashai, A. S. (2008a) A new neural network based model for hysteretic behavior of materials, *International Journal for Numerical Methods in Engineering*, 73(4): 447–469.

Yun, G. J., Ghaboussi, J. and Elnashai, A. S. (2008b) Self-learning simulation method for inverse nonlinear modeling of cyclic behavior of connections, *Computer Methods in Applied Mechanics and Engineering*, 197(33–40): 2836–2857.

Zhang, M. (1996) Determination of neural network material models from structural tests, PhD thesis, Department of Civil and Environmental Engineering, University of Illinois at Urbana-Champaign, Urbana, IL.

Index